# Lecture Notes of the Institute for Computer Sciences, Social-Informatics and Telecommunications Engineering 27

Patty Kostkova (Ed.)

# Electronic Healthcare

Second International ICST Conference, eHealth 2009
Istanbul, Turkey, September 23-15, 2009
Revised Selected Papers

Volume Editor

Patty Kostkova
The City University
City ehealth Research Center (CeRC)
School of Community and Health Sciences
Northhampton Square
EC1V 0HB London, United Kingdom
E-mail: patty@soi.city.ac.uk

Library of Congress Control Number: 2009943614

CR Subject Classification (1998): J.3, K.4, H.2.8, H.3.3, I.2.6, J.1, K.3

ISSN 1867-8211
ISBN-10 3-642-11744-9 Springer Berlin Heidelberg New York
ISBN-13 978-3-642-11744-2 Springer Berlin Heidelberg New York



springer.com


Printed in Germany

Typesetting: Camera-ready by author, data conversion by Scientific Publishing Services, Chennai, India
Printed on acid-free paper SPIN: 12985005 06/3180 5 4 3 2 1 0

# Preface

It is my great pleasure to introduce this special issue of LNSV comprising the scientific publications presented at ehealth 2009: The second Congress on Electronic Healthcare for the 21st Century, which took place in Istanbul, Turkey during September 23–25, 2009.

Building on the first ehealth 2008 congress held in London, UK, the key topic of ehealth 2009 was investigating a realistic potential of the Internet in providing evidence-based healthcare information and education to patients and global users. The proudly defined aim of ehealth 2009 — bringing together the three medical sectors: academia, industry and global healthcare institutions — was met and made the congress a truly unique event. The formal and informal discussions among the conference participants led to numerous stimuli for new collaborations.

We accepted 26 full and 10 short technical presentations by speakers from all over the world, having received over 80 submissions. In addition to two keynotes, the commercial angle was provided by invited industrial speakers representing a wide range of healthcare IT companies including Corinne Marsolier of Cisco, Glenn Kenneth Bruun (CSAM Health), Luis Falcón (Thymbra) and Johan Muskens (Philips Research Europe), as well as international healthcare organizations such as Med-e-Tel represented by the international coordinator Frederic Lievens. The industry presentations were complemented by real-world healthcare delivery expertise presented by the delegation from the European Centre for Disease Prevention and Control (ECDC), Stockholm, led by Leonora Brooke and László Balkányi, who presented two technical papers and had a dedicated session to demonstrate the wide spectrum of their ICT activities.

Moreover, ehealth 2009 had the honor of enjoying presentations from two prestigious keynote speakers delivering contrasting and very stimulating talks: Harald Korb from Vitaphone GmbH described the need for telemedicine in his keynote titled: "Telemedicine Is Essential to National Health Strategies in the Management of Chronic Diseases — The Example of Cardiology," and Erik van der Goot from the EC Joint Research Centre, Italy, presented the media-monitoring service for infectious disease outbreak detection run by the JRC: "The Medical Information System MedISys: A Tool for Medical Intelligence."

Julius Weinberg, the Acting Vice Chancellor of the City University, skilfully moderated the panel discussion "Challenges of eHealth in the 21st Century: When Will Industry and Academia Meet the Needs of Healthcare Delivery?" that generated a vibrant and heated discussion. A key topic mentioned several times by the panellists and the audience was the need to bridge the existing gap between IT-driven healthcare development and the real needs of medical professionals and the healthcare sector. Ebru Başak Aköz, National ICT Expert for Turkey, EU Framework Programmes, gave a brief introduction to the FP7 funding schemes.

The technical sessions ran in parallel and covered a wide range of subjects including topical ideas such as the use of Web 2.0 tools for epidemic intelligence, security and patient safety, support for elderly citizens as well as traditional knowledge management issues such as medical ontologies and terminology services. Above all, ehealth 2009 also contributed to closing the digital divide by presenting the use of telemedicine for health promotion and disease prevention in emerging economies in South America.

The Poster Session, chaired by Jason Bonander of CDC, provided a unique opportunity to present work-in-progress projects and the very popular Demonstration Session, chaired by Steve Bunting of City University, offered a hands-on experience in an informal setting.

Organized and chaired by Steve Bunting on the first evening of the conference, the novel activity "Speed Dating," encouraging industry–academia–healthcare networking, was a great success enabling participants to get to know each other in an enjoyable and informal setting.

I would like to thank everyone who contributed to the success of ehealth 2009. The authors of all submitted papers, the speakers, the invited and the keynote presenters, the Program Committee members, reviewers and Session Chairs and above all the Local Chair, Aslı Uyar, together with student volunteers and local organizers. Finally, we thank the ICST and the Create-Net for sponsoring the event, Springer for publishing this LNSV book and the City University for supporting ehealth 2009 by allowing me to chair the event and my five colleagues to accept other roles and assist in the organization.

October 2009 Patty Kostkova

# Organization

## Steering Committee Chair

Imrich Chlamtac — President, Create-Net Research Consortium

## Steering Committee Member

Tibor Kovacs — Director of Business and Technology Affairs, ICST

## General and Scientific Chair

Patty Kostkova — City eHealth Research Centre, City University, UK

## Programme Co-chair

Muttukrishnan Rajarajan — Mobile Networks Research Centre, City University, UK

## Knowledge Transfer Co-chair

Steve Bunting — Knowledge Transfer, City University, UK

## Industry Co-chair

Simon Thompson — BT, UK

## Local Chair

Asli Uyar — Bogazici University, Istanbul, Turkey

## Web Chair

Helen Oliver — City eHealth Research Centre, City University, UK

## Poster Chair

Jason Bonander — Director of Knowledge Management, CDC, Atlanta, USA

## Technical Program Committee

| | |
|---|---|
| Anne Adams | The Institute of Educational Technology, The Open University, UK |
| Oliver Amft | Signal Processing Systems, Eindhoven University of Technology, The Netherlands |
| Elske Ammenwerth | Director of the Institute for Health Information Systems, UMIT, Austria |
| Simon Attfield | University College London Interaction Centre (UCLIC), UCL, UK |
| László Balkányi | Knowledge Manager, ECDC, Sweden |
| Ayşe Bener | Bogazici University, Turkey |
| Ann Blandford | University College London Interaction Centre (UCLIC), UCL, UK |
| Jason Bonander | CDC, USA |
| Arnold Bosman | ECDC, Sweden |
| Steve Bunting | Knowledge Transfer, City University, UK |
| Albert Burger | Heriot-Watt University, UK |
| Juan Chia | Worldwide Clinical Trials, UK |
| Nadir Ciray | Bahceci Women Health Care Centre, Istanbul, Turkey |
| Ulises Cortés | Universitat Politècnica de Catalunya, Spain |
| Wendy Currie | University of Warwick, UK |
| Kay H. Connelly | Indiana University, USA |
| Ed de Quincey | City eHealth Research Centre, City University, UK |
| Gayo Diallo | Laboratory of Applied Computer Science LISI-ENSMA (FUTUROSCOPE Poitiers), France |
| Claudio Eccher | FBK-Center for Scientific and Technological Research, Trento, Italy |
| Jonathan Elford | Department of Public Health, City University, London, UK |
| Henrik Eriksson | Linköping University, Linköping, Sweden |
| Bob Fields | Middlesex University, UK |
| Victor M. Gonzalez | University of Manchester, UK |
| Floriana Grasso | University of Liverpool, UK |
| Femida Gwadry-Sridhar | London Health Sciences Centre, University of Western Ontario, Canada |
| David Hansen | CSIRO Australian E-Health Research Centre, Brisbane, Australia |
| Gawesh Jawaheer | City eHealth Research Centre, City University, UK |
| Simon Jupp | Bio-Health Informatics Group, The University of Manchester, UK |
| Khaled Khelif | Project Edelweiss, INRIA Sophia Antipolis Méditerrannée, France |
| Patty Kostkova | City eHealth Research Centre, City University, UK |
| Panayiotis Kyriacou | City University, UK |
| Shirley Large | NHS Direct, Hampshire, UK |

| Cecil Lynch | UC Davis School of Medicine, California, USA |
|---|---|
| Maria G. Martini | Kingston University, UK |
| Kenneth McLeod | Heriot-Watt University, UK |
| Antonio Moreno | Universitat Rovira i Virgili, Spain |
| Paddy Nixon | University College Dublin, Ireland |
| Helen Oliver | City eHealth Research Centre, City University, UK |
| Venet Osmani | Multimedia, Interaction and Smart Environments (MISE), CREATE-NET, Trento, Italy |
| Vladimir Prikazsky | ECDC, Stockholm, Sweden |
| Rob Procter | University of Edinburgh, UK |
| Veselin Radokevic | Information Engineering Research Centre, City University, UK |
| Muttukrishnan Rajarajan | Mobile Networks Research Centre, City University, UK |
| Dietrich Rebholz-Schuhmann | The Rebholz Group, Cambridge, UK |
| David Riaño | Universitat Rovira i Virgili, Spain |
| Heiko Schuldt | Datenbanken und Informationssysteme, University of Basel, Switzerland |
| Tacha Serif | Yeditepe University, Turkey |
| Crystal Sharp | Lawson Health Research Institute, UK |
| Simon Thompson | BT, UK |
| Tammy Toscos | Indiana University, USA |
| Aslı Uyar | Bogazici University, Turkey |
| Jan Vejvalka | Charles University, Czech Republic |
| Dasun Weerasinghe | City University, UK |
| Yeliz Yesilada | Human Centered Web Lab, The University of Manchester, UK |
| Jana Zvarova | EuroMISE, The Academy of Sciences of CR, Czech Republic |

# Table of Contents

## Session 1: Telehealth and Mobile Health Solutions



## Session 2: Outbreak Management, Web 2.0 and Public Health Communication



## Session 3: EPR: Trust, Security and Decision Support

## Session 4: ICT Support for Patients and Healthcare Organizations



## Session 5: Evaluation of ICT in Healthcare

## Session 6: Healthcare Knowledge Management and Ontologies



## Session 7: Web 2.0, Multimedia and Personalisation

## Session 8: eHealth Automation and Decision Support



## Session 9: European Centre for Disease Prevention and Control

# Model Checking for Robotic Guided Surgery

H. Mönnich, J. Raczkowsky, and H. Wörn

IPR Karlsruhe, University of Karlsruhe, Karlsruhe, Germany
{moennich,rkowsky,wörn}@ira.uka.de

**Abstract.** This paper describes a model checking approach for robotic guided surgical interventions. The execution plan is modeled with a workflow editor as a petri net. The net is then analyzed for correct structure and syntax with XMLSchema. Petri nets allow checking for specific constraints, like soundness. Still the possibility to prove the net with runtime variables is missing. For this reason model checking is introduced to the architecture. The Petri-Net is transformed to the Model Checking language of NuSMV2, an open source model checking tool. Conditions are modeled with temporal logic and these specifications are proved with the model checker. This results in the possibility to prove the correct initialization of hardware devices and to find possible runtime errors. The workflow editor and model checking capabilities are developed for a demonstrator consisting of a KUKA lightweight robot, a laser distance sensor and ART tracking for CO2 laser ablation on bone.

**Keywords:** Workflow Systems, service oriented architecture, CORBA, knowledge management, temporal logic.

## 1 Introduction

One important aspect of the AccuRobAs project is to provide a modular system architecture for robot assisted surgical interventions. This goal of a modular system is hard to reach combined with the requirements of surgical applications like safety and hard realtime environments. The realtime problem has to be solved with special protocols and software environments. The safety problem is more complex and cannot be solved in general. The reason is the still unsolved problem of understanding a language and the underlying semantic as pointed out by[1][2]. But different approaches to solve this problem partially have been developed. One example is model checking that is successfully used for circuit design. Model checking is the process of checking whether a given structure is a model of a given logical formula. The model is often simplified. The concept is general and applies to all kinds of logics and suitable structures. A simple model-checking problem is testing whether a given formula in the propositional logic is satisfied by a given structure. The main drawback of model checkers is the state explosion problem that is heavily limiting the usage of these tools. Model checking was already the interest of the NASA project PLEXIL[3]. Plexil offers a special programming language for execution plans. These execution plans can then be proved by a model checker. PLEXIL uses the LTSA model checker [4]. This already showed the successful usage of a model checking tool for a planning

P. Kostkova (Ed.): eHealth 2009, LNICST 27, pp. 1–4, 2010.

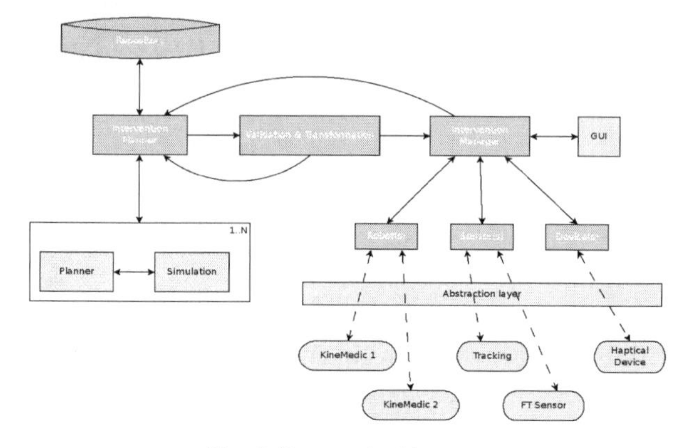


**Fig. 1.** System Architecture

language. The main drawback is the usage of a special planning language that is not intuitive for the user. Workflows are more common to field of medical applications. A workflow formalism that is well proven is the petri net workflow as described by van der Aalst et al.[5]. A petri net has formal semantics with a clear and precise definition. Several enhancements for classical petri nets have been developed and defined formally like colored, timed or hierarchical petri nets. Twenty different workflow patterns were discovered for workflow nets that are all supported by petri nets, even the more rarely used implicit or operation [6]. For this reason they were chosen for the AccuRobAs project for the surgical application field. In figure 1 the system architecture is shown. In this paper the Validation and Transformation step is described.

## 2 Transformation and Validation

The validation and transformation step is important to ensure the correct behaviour of the plan. Validation can be divided into two steps. The first step is to check the syntax and the corresponding structure of a plan. This part is well known from constructing compilers in computer science and can be easily adapted to the plans with XMLSchema. The second step is more challenging. To ensure the correct behavior of programs a model checker is used to solve this problem partially. The last step is the transformation into a state machine for real-time execution. The workflow plan is transferred into a state machine that is executed by the Intervention Manager, currently an implementation based on SCXML Apache.

### 2.1 Model Checking

The software NuSMV [7] is used here as a model checker. NuSMV is an updated version of the SMV symbolic model checker. NuSMV has been developed as a joint project between ITC-IRST (Istituto Trentino di Cultura, Istituto per la Ricerca Scientifica e Tecnologica in Trento, Italy), Carnegie Mellon University, the University of Genoa and the University of Trento. It is a reimplementation and extension of SMV, the first model checker based on Binary Decision Diagrams (BDDs).

The transformation step from a petri net to the model checker NuSMV is the following: 1.) the workflow is constructed inside the yawl workflow editor with all conditions 2.) the yawl workflow is exported to an xml file 3.) the transformation from the xml petri net to the NuSMV language as an text file is done with an xslt stylesheet 4.) NuSMV2 is evaluating the specifications. A similar transformation has been described by Schaad and Sohr[8].

The transformation itself is performed in the following way: Every task and condition is modeled as a variable inside the NuSMV2 language. The variable has two possible values: "wait" and "run". They indicate the current status of the state.

## 3 Model Checking Examples

Figure 2 is shows a simple example workflow. Here a robotic device, called LBR, is used to move to a specific starting position and to measure the distance between TCP point and the target with a laser distance sensor. Under all circumstances it should be possible to reach a safe state and to move the robot away from the patient. In this case the following specifications could be checked: 1.) Is the state "safePos" reachable, 2.) the state "safePos" should never be activated if the condition is not true and 3.) if the safePos is desired, the next step is always the safe state, excepting the two cases for the start and end state.

Reachability: `SPEC EF (safePos_8 = runs)`
Safety: `SPEC AG (!goToSafePos -> AX safePos_8 != runs)`
Liveness: `SPEC AG (goToSafePos & InputCondition_1 != runs & Output-Condition_2 != runs) -> AX safePos_8 = runs`

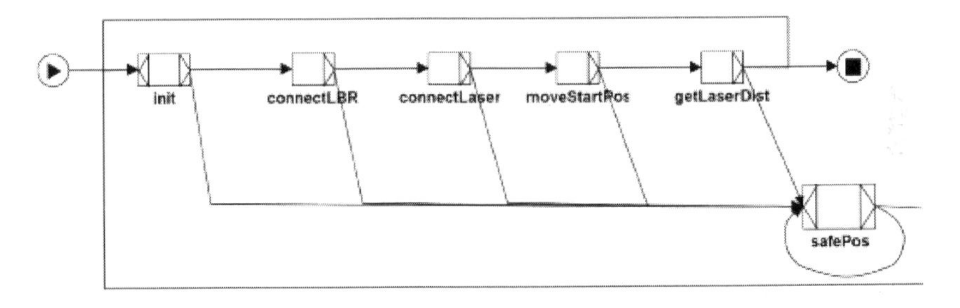


**Fig. 2.** Simple example workflow

## 4 Results

In this work an approach for model checking for surgical interventions is presented. A method is described how to prove constraints inside the workflow. The main problem however remains, it can only be shown that a workflow satisfies a given specification or not. It can be shown that a workflow is not violating against a specification but not that the workflow behaves as expected.

## 5 Future Works

The current conversion from petri net to the NuSMV2 model checker is still very limited and should be compared to other methods. Even the specifications must be inserted by the user after the conversion process. This should be done automatically to show the full potential of this approach. It would be helpful to develop a set of typical rules that must be satisfied for any kind of surgical procedure.

## Acknowledgement

This work has been funded by the European Commission's Sixth Framework Program within the project Accurate Robot Assistant - ACCUROBAS under grant no. 045201.

The author would like to thank Henrik Keller who implemented the complete transformation and developed the example during his student research project [9] at IPR Karlsruhe.

## References

1. Wittgenstein, L.: Philosophische Untersuchungen, Herausgegeben von Joachim Schulte. Wissenschaftliche Buchgesellschaft. Frankfurt am Main (2001)
2. Searle, J.R.: Minds, brains, and programs. Behavioral and Brain Sciences 3 (1980)
3. Siminiceanu, R.: Model Checking Abstract PLEXIL Programs with SMART. NASA CR-2007-214542 (April 2007)
4. Magee, J., Kramer, J.: Concurrency. In: State Models and Java Programs.: State Models and Java Programs. Wiley & Sons, Chichester (1999)
5. ter Hofstede, A.H., van der Aalst, W.M.: YAWL: yet another workflow language. Information Systems 30(4), 245–275 (2005)
6. Russell, N., ter Hofstede, A.H.M., van der Aalst, W.M.P., Mulyar, N.: Workflow Control-Flow Patterns. A Revised View
7. Cimatti, E.M., Clarke, E., Giunchiglia, F., Giunchiglia, M., Pistore, M., Roveri, R.: NuSMV 2: An OpenSource Tool for Symbolic Model Checking. In: Brinksma, E., Larsen, K.G. (eds.) CAV 2002. LNCS, vol. 2404, p. 359. Springer, Heidelberg (2002)
8. Schaad, A., Sohr, K.: A workflow instance-based model-checking approach to analysing organisational controls in a loan origination process. In: 1st International Workshop on Secure Information Systems (SIS 2006), Wisla, Polen (2006)
9. Keller, H.: Erhöhung der Sicherheit von Robotersteuerungen in der Medizin durch Model Checking, University of Karlsruhe (2008)

# Intelligent Mobile Health Monitoring System (IMHMS)

Rifat Shahriyar, Md. Faizul Bari, Gourab Kundu, Sheikh Iqbal Ahamed, and Md. Mostofa Akbar

Department of Computer Science and Engineering,
Bangladesh University of Engineering and Technology, Bangladesh
{rifat1816,faizulbari,gourab.kundu08}@gmail.com,
sheikh.ahamed@mu.edu, mostofa@cse.buet.ac.bd

**Abstract.** Health monitoring is repeatedly mentioned as one of the main application areas for Pervasive computing. Mobile Health Care is the integration of mobile computing and health monitoring. It is the application of mobile computing technologies for improving communication among patients, physicians, and other health care workers. As mobile devices have become an inseparable part of our life it can integrate health care more seamlessly to our everyday life. It enables the delivery of accurate medical information anytime anywhere by means of mobile devices. Recent technological advances in sensors, low-power integrated circuits, and wireless communications have enabled the design of low-cost, miniature, lightweight and intelligent bio-sensor nodes. These nodes, capable of sensing, processing, and communicating one or more vital signs, can be seamlessly integrated into wireless personal or body area networks for mobile health monitoring. In this paper we present Intelligent Mobile Health Monitoring System (IMHMS), which can provide medical feedback to the patients through mobile devices based on the biomedical and environmental data collected by deployed sensors.

**Keywords:** Mobile Health care, Health Monitoring System, Intelligent Medical Server.

## 1 Introduction

Pervasive computing is the concept that incorporates computation in our working and living environment in such a way so that the interaction between human and computational devices such as mobile devices or computers becomes extremely natural and the user can get multiple types of data in a totally transparent manner. The potential for pervasive computing is evident in almost every aspect of our lives including the hospital, emergency and critical situations, industry, education, or the hostile battlefield. The use of this technology in the field of health and wellness is known as pervasive health care. Mobile computing describes a new class of mobile computing devices which are becoming omnipresent in everyday life. Handhelds, phones and manifold embedded systems make information access easily available for everyone from anywhere at any time. We termed the integration of mobile computing to pervasive health care as mobile health care. The goal of mobile health care is to provide health care services to anyone at anytime, overcoming the constraints of place, time

P. Kostkova (Ed.): eHealth 2009, LNICST 27, pp. 5–12, 2010.

and character. Mobile health care takes steps to design, develop and evaluate mobile technologies that help citizens participate more closely in their own health care. In many situations people have medical issues which are known to them but are unwilling or unable to reliably go to a physician. Obesity, high blood pressure, irregular heartbeat, or diabetes is examples of such common health problems. In these cases, people are usually advised to periodically visit their doctors for routine medical checkups. But if we can provide them with a smarter and more personalized means through which they can get medical feedback, it will save their valuable time, satisfy their desire for personal control over their own health, and lower the cost of long term medical care. A number of bio-sensors that monitor vital signs, environmental sensors (temperature, humidity, and light), and a location sensor can all be integrated into a Wearable Wireless Body/Personal Area Network (WBAN/WPAN). This type of networks consisting of inexpensive, lightweight, and miniature sensors can allow long-term, unobtrusive, ambulatory health monitoring with instantaneous feedback to the user about the current health status and real-time or near real-time updates of the user's medical records. Such a system can be used for mobile or computer supervised rehabilitation for various conditions, and even early detection of medical conditions. When integrated into a broader tele-medical system with patients' medical records, it promises a revolution in medical research through data mining of all gathered information. The large amount of collected physiological data will allow quantitative analysis of various conditions and patterns. Researchers will be able to quantify the contribution of each parameter to a given condition and explore synergy between different parameters, if an adequate number of patients are studied in this manner.

In this paper we present a bio-sensor based mobile health monitoring system named as "Intelligent Mobile Health Monitoring System (IMHMS)" that uses the Wearable Wireless Body/Personal Area Network for collecting data from patients, mining the data, intelligently predicts patient's health status and provides feedback to patients through their mobile devices. The patients will participate in the health care process by their mobile devices and thus can access their health information from anywhere any time. Moreover, so far there is no automated medical server used in any of the work related to mobile health care. To maintain the server a large number of specialist are needed for continuous monitoring. The presence of a large number of specialists is not always possible. Moreover in the third world countries like ours specialist without proper knowledge may provide incorrect prescription. That motivates us to work for an intelligent medical server for mobile health care applications that will aid the specialists in the health care. As a large amount of medical data is handled by the server, the server will perform mine and analyze the data. With the result of mining, analysis and suggestions and information provided by the specialists in the critical scenarios the server can learn to provide feedback automatically. Moreover as time grows the server will trained automatically by mining and analyzing data of all the possible health care scenarios and become a real intelligent one. Our main contribution here is the Intelligent Medical Server which is a novel idea in the field of mobile health care. The outline of this paper is as follows: System Architecture of IMHMS is described in Section 2 followed by the evaluation of IMHMS in Section 3. Our future research direction and concluding remarks are in Section 4.

## 2 System Architecture

IMHMS collects patient's physiological data through the bio-sensors. The data is aggregated in the sensor network and a summary of the collected data is transmitted to a patient's personal computer or cell phone/PDA. These devices forward data to the medical server for analysis. After the data is analyzed, the medical server provides feedback to the patient's personal computer or cell phone/PDA. The patients can take necessary actions depending on the feedback. The IMHMS contains three components. They are: 1. *Wearable Body Sensor Network [WBSN]*, 2. *Patient's Personal Home Server [PPHS]* and 3. *Intelligent Medical Server [IMS]*.

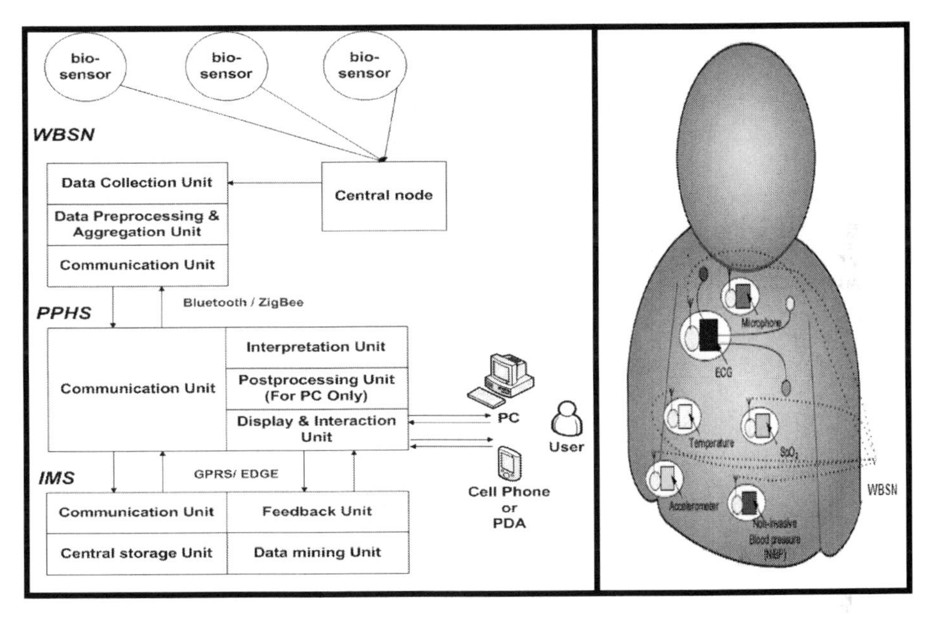


**Fig. 1.** System Architecture

**Fig. 2.** WBSN

They are described below.

***Wearable Body Sensor Network [WBSN]***

Wearable Body Sensor Network is formed with the wearable or implantable bio-sensors in patient's body. These sensors collect necessary readings from patient's body. For each organ there will be a group of sensors which will send their readings to the group leader. The group leaders can communicate with each others. They send the aggregated information to the central controller. The central controller is responsible for transmitting patient's data to the personal computer or cell phone/PDA. A recent work suggested that for wireless communication inside the human body, the tissue medium acts as a channel through which the information is sent as electromagnetic (EM) radio frequency (RF). So in WBSN, information is transmitted as electromagnetic (EM) radio frequency (RF) waves. The central controller of the WBSN communicates with the Patients Personal Home Server [PPHS] using any of the three wireless protocols: Bluetooth, WLAN (802.11) or ZigBee. Bluetooth can be used for short range distances between the central controller and PPHS. WLAN can be used to support more distance between them. Now days ZigBee introduces itself as a specialized wireless protocol suitable for pervasive and ubiquitous applications. So ZigBee can be used for the communication too. As we all know that Bluetooth, WLAN and ZigBee have very different characteristics and they should be analyzed for which one makes most sense for use in our system. Currently we are working on it by comparative studies and analysis to choose the best one.

***Patient's Personal Home Server [PPHS]***
The patient's personal home server can be a personal computer or mobile devices such as cell phone/PDA. We suggest mobile devices because it will be more suitable for the users to use their mobile devices for this purpose. PPHS collects information from the central controller of the WBSN. PPHS sends information to the Intelligent Medical Server [IMS].PPHS contains logics in order to determine whether to send the information to the IMS or not. Personal Computer based PPHS communicates with the IMS using Internet. Mobile devices based PPHS communicates with the IMS using GPRS / Edge / SMS. The best way to implement IMS is by Web Service or Servlet based architecture. The IMS will act as the service provider and the patients PPHS will act as the service requester. By providing these types of architecture, a large number of heterogeneous environments can be supported with security. So personal computer or cell phone/PDA can be connected easily to a single IMS without any problem.

***Intelligent Medical Server [IMS]***
Intelligent Medical Server [IMS] receives data from all the PPHS. It is the backbone of this entire architecture. It is capable of learning patient specific thresholds. It can learn from previous treatment records of a patient. Whenever a doctor or specialist examines a patient, the examination and treatment results are stored in the central database. IMS mines these data by using state-of-the-art data mining techniques such as neural nets, association rules, decision trees depending on the nature and distribution of the data. After processing the information it provides feedback to the PPHS or informs medical authority in critical situations. PPHS displays the feedback to the patients. Medical authority can take necessary measures. The IMS keeps patient specific records. It can infer any trend of diseases for patient, family even locality. IMS can cope with health variations due to seasonal changes, epidemics etc. IMS is controlled and monitored mainly by specialized physicians. But even a patient can help train IMS by providing information specific to him. After mining the database stored in IMS, important information regarding general health of the people can be obtained. It can help the authority to decide health policies. Large numbers of patients will be connected to the IMS using their PPHS. So security of the patients is a major issue here. So RFID can be used in this purpose. Radio-frequency identification (RFID) is an automatic identification method, relying on storing and remotely retrieving data using devices called RFID tags or transponders. An RFID tag is an object that can be applied to or incorporated into a product, animal, or person for the purpose of identification using radio waves. Some tags can be read from several meters away and beyond the line of sight of the reader. So security can be provided by providing RFID tags to each patient. Our main contribution is the Intelligent Medical Server (IMS) which is a novel idea. So we are describing it in more details with possible scenarios below. ***In intensive care units, there are provisions for continuously monitoring***

***patients.*** Their heart rates, temperatures etc. are continuously monitored. But in many cases, patients get well and come back to home from hospital. But the disease may return, he may get infected with a new disease, there may be a sudden attack that may cause his death. So in many cases, patients are released from hospital but still they are strongly advised to be under rest and observation for some period of time (from several days to several months). In these cases, IMHMS can be quite handy. ***Patients of blood pressure frequently get victimized because of sudden change of pressures.*** It cannot be foreseen and also a normal person cannot be kept under medical observation of a doctor or a hospital all days of a year. Blood pressures change suddenly and can be life-treating. Using IMHMS, they can get alerts when their blood pressure just starts to become high or low. ***Patient's data (temperature, heart rate, glucose level, blood pressure etc.) will be frequently measured and sent to PPHS.*** Period of sending (say every 3 min) can be set from the patient in the central controller of WBSN. Normally glucose level will be sent after several days or a week. Heart rates can be sent every minute and temperatures can be sent after half an hour etc. But these can be parameterized to ensure that when a patient is normal, not many readings will be sent so that sensors have a longer life-time. But when the patient is ill, readings will be taken frequently and sent to PPHS. ***PPHS will have some logic to decide whether the information is worthy of sending to IMS***. Say, temperature is in safety range(less than 98F), and then PPHS will not send this info to IMS to save cost for the patient. Again say, glucose level is safe and same as the last several days, then this info also need not be sent. Data must be sent to IMS when there is a change in status (say temperature of the patient goes to 100F from 98F or a patient with severe fever 102F has just got temperature down to 99F). Again if there is a sudden change in blood pressure or glucose level, then this info must be sent to IMS. ***IMS learns patient specific threshold.*** Say the regular body temperature of a patient is 98.2F whereas one person feels feverish if his body temperature is 98.2F. By employing an averaging technique over a relatively long time, IMS can learn these thresholds for patients. However, patients can also give these thresholds as inputs based on directions of their doctors. ***Using IMS, one can view his medical history date wise, event wise etc.*** IMS can perform data mining on a particular patient data to discover important facts. Suppose a person has medium high temperature that starts at evening and lasts till midnight. If this phenomenon continues for several days, IMS will automatically detect this fact and send a message to PPHS saying "You frequently have short-period fever that may be a symptom of a bad disease. Consult doctor immediately". ***Using IMS, one can view his medical history date wise, event wise etc.*** A patient can also enter extra information like he has had chest pain today, or he is frequently vomiting, he has rashes on body etc. in PPHS. In IMS, there will be a set of rules for preliminary prediction of disease. These rules will be pre learnt based on neural network or data mining of existing disease databases that are available over web. Now IMS, with the additional information, will check the rules. If it finds a matching rule, then it will predict the disease and send the message to PPHS. ***The most important fact about IMS is that it can help stop the spread of diseases.*** Whenever it finds that several people from same locality over a small period of time are having the same disease, it will predict that the disease is spreading out in that locality so that authority can take immediate action. Say, when some people of the same area report that they are having high fever, pain over body and rashes, IMS will report this which the doctors can interpret that dengue is breaking out in that area and the authority has a chance to take actions at the very first stage so that epidemic can be avoided.

## 3 Evaluation

To evaluate IMHMS, we implement a prototype of different components of IMHMS.

We are working on building WBSN. We consider the data provided by the bio-sensors as a well structured XML file. A sample XML file is shown in Figure 3 where a patient's Temperature, Glucose-level and Blood-pressure are measured continuously over a period of time. Two possible implementations are there for PPHS. It can be implemented in personal computer. While implementing for personal computer, the most suitable communication media between WBSN and PPHS is Bluetooth because of its availability and low cost. The personal computer based PPHS implementation required Bluetooth Server setup in the personal computer. The medical data of the patients will be transferred from WBSN to PPHS through the Bluetooth Server. Then the personal computer based PPHS processes the data and send necessary data to the IMS. But we suggest mobile devices for implementing PPHS because it will be more suitable for the users to use their cell phones or PDA in this purpose. The real mobility of the solution can be provided by mobile devices. We choose J2ME based custom application so that it can be deployed immediately in a large number of available cell phones or PDA available in the market. The J2ME based PPHS automatically collect patient's data from the WBSN and transfer it to the IMS. It is also responsible for displaying results and feedback from the IMS to any specific patients. We implemented the skeleton of the IMS. IMS is built with the Java Servlet based architecture. To connect to the IMS, PPHS requires software to be installed. We implemented a J2ME application that processes the XML file of patient's data using KXML which is an open source XML parser. The application connects to the IMS using GPRS or EDGE. It can connect using SMS also if SMS receiving capable application can be developed in the IMS. Our J2ME application connects to the IMS's Web Servlet by GPRS or EDGE. To implement the SMS based portion the IMS must be interfaced with a number of cell phones or PDA in order to receive SMS from the PPHS and send the feedback to the PPHS as SMS. The flow diagram of the implementation is shown in the Figure 4. The WBSN collects patient data and send the data to the PPHS. PPSH receives the data and processed the data to reduce the transmission of unnecessary data to the IMS. The PPHS communicates with the IMS using GPRS or EDGE. The IMS contains a Data Mining Unit, a Feedback Unit and a central database. The database contains the entire patients' profile, continuous health data and a large set of rules for data mining operations. The Data Mining unit processes the data and returns the feedbacks and results to the Feedback Unit. The feedback unit then sends the data to the corresponding PPHS. Moreover the patient's can login to the IMS using authorized patient-id and password to provide information manually and to view the patient's entire history. Some screenshots of these activities are shown in the figure. Figure 5.1 and 5.2 show the interface in IMS for patients profile information and manual health data submission. Figure 5.3 shows one patient's entire medical history with the feedbacks and results stored in the IMS's central database. Figure 5.4 and 5.5 show the automated health data collection of J2ME based PPHS and display of feedbacks provided by IMS based on the collected data.

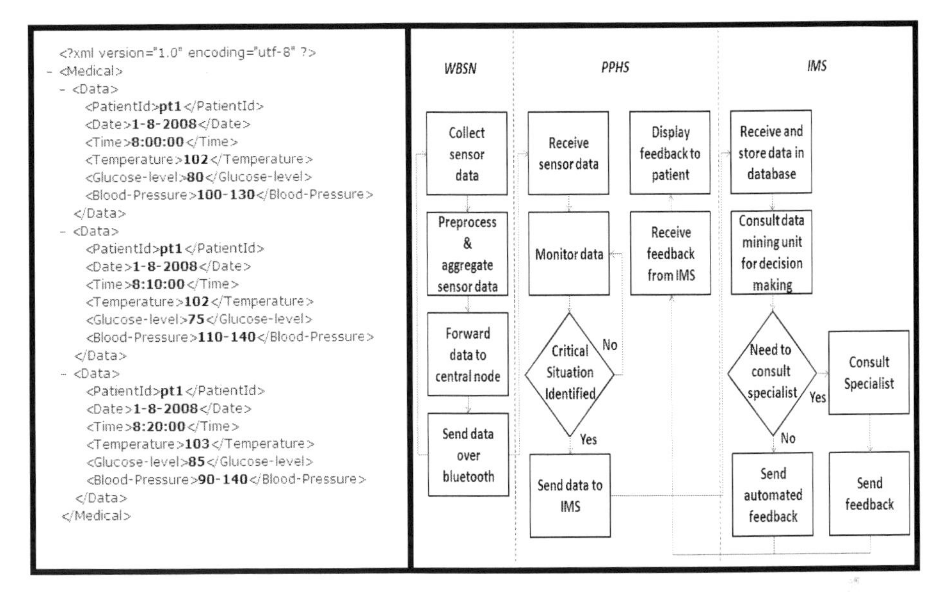

**Fig. 3.** Patient's Health Data

**Fig. 4.** Flow Diagram

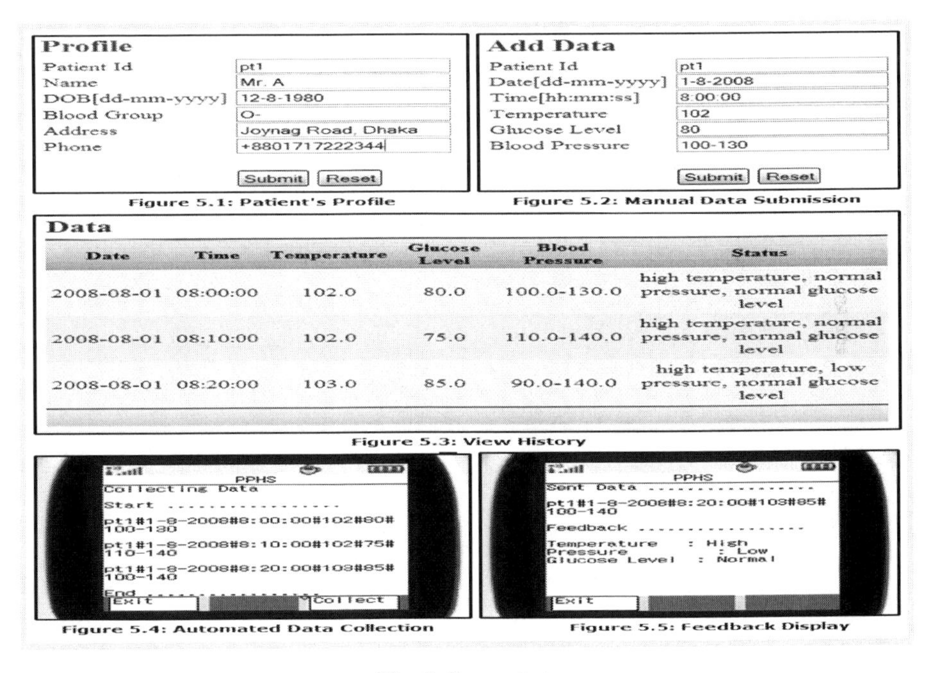


**Fig. 5.** Screenshots

## 4 Future Works and Conclusion

The whole system of mobile health care using biosensor network places forward some future works such as finding the most effective mechanism for ensuring security in bio-sensors considering the severe restrictions of memory and energy, representing the collected data in the most informative manner with minimal storage and user interaction, modeling of data so that the system will not represent all the data but only relevant information thus saving memory. These are the generic works that can be done in future in the sector of mobile health care. For IMHMS our vision is much wider. We think of a system where the patients need not to do any actions at all. With the advancement of sensor technologies it is not far enough when the bio-sensors itself can take necessary actions. A patient needed glucose does not need to take it manually rather the bio-sensors can push the glucose to the patient's body depending on the feedback from the IMS. It seems to be impossible to achieve by everybody. But nothing is impossible. Today we imagine of something and see that it is implemented in the near future. But if we stop imagine and thinking then how impossible can be made possible? This paper demonstrates an intelligent system for mobile health monitoring. Smart sensors offer the promise of significant advances in medical treatment. As the world population increases, the demand for such system will only increase. We are implementing the IMHMS to help the individuals as well as the whole humanity. Our goals will be fulfilled if the IMHMS can help a single individual by monitoring his or her health and cautions him to take necessary actions against any upcoming serious diseases.

## References

1. Ahamed, S.I., Haque, M.M., Stamm, K., Khan, A.J.: Wellness assistant: A virtual wellness assistant using pervasive computing. In: ACM Symposium on Applied Computing (SAC), Seoul, Korea, March 2007, pp. 782–787 (2007)
2. Gupta, S.K.S., Lalwani, S., Prakash, Y., Elsharawy, E., Schwiebert, L.: Towards a propagation model for wireless biomedical applications. In: IEEE International Conference on Communications (ICC), May 2003, vol. 3, pp. 1993–1997 (2003)
3. Korhonen, I., Lappalainen, R., Tuomisto, T., Koobi, T., Pentikainen, V., Tuomisto, M., Turjanmaa, V.: Terva: wellness monitoring system. In: 20th Annual International Conference of the IEEE, Engineering in Medicine and Biology Society, October 1998, vol. 4(29), pp. 1988–1991 (1998)
4. Milenkovic, A., Otto, C., Jovanov, E.: Wireless sensor networks for personal health monitoring: Issues and an implementation. Computer Communications (Special issue: Wireless Sensor Networks: Performance, Reliability, Security, and Beyond) 29(13-14), 2521–2533 (2006)
5. Parkka, J., van Gils, M., Tuomisto, T., Lappalainen, R., Korhonen, I.: Wireless wellness monitor for personal weight management. In: IEEE EMBS International Conference on Information Technology Applications in Biomedicine, November 2000, pp. 83–88 (2000)
6. Varshney, U.: Pervasive healthcare and wireless health monitoring. Journal on Mobile Networks and Applications (Special Issue on Pervasive Healthcare) 12(2-3), 111–228 (2007)

# Mobile Health Access for Diabetics in Rural Areas of Turkey – Results of a Survey

Emine Seker and Marco Savini

University of Fribourg, Boulevard de Pérolles 90, 1701 Fribourg, Switzerland
{emine.seker,marco.savini}@unifr.ch

**Abstract.** Extending the reach of medical professionals in rural areas is one of the goals using mobile health technologies. This paper illustrates the results of a survey conducted in 2008 in Turkey asking medical professionals about their current ICT usage and opinions about using mobile technologies in order to help patients with diabetes. The goal is to reduce the information gap between patients and medical professionals by allowing sending the information electronically using mobile technologies. This will improve both the interaction between various actors and also improve the treatment, as important trends of this chronic disease can be discovered on time.

**Keywords:** mobile health, diabetes, turkey, survey, rural area, mhealth, ehealth.

## 1 Mobile Devices in eHealth

A very large percentage of the European population owns at least one mobile device, typically a mobile phone. Their ubiquity, connectivity and increasing features are reasons for their use in the electronic health sector.

A number of researchers have worked on the idea of assigning mobile devices to patients. Furthermore, the WHO interprets the high demand of non-OECD countries for telemedicine and the use of remote medical expertise as the need to improve the available health care resources in less developed areas [1]. It is possible to discriminate between the following three domains of mobile health applications:

1. Mobile devices are used to help the patient by providing information.
2. Mobile devices are used to transmit physiological parameters.
3. Mobile devices are used to alert patients or medical professionals when certain physiological parameters become critical.

### 1.1 Providing Information to Patients

The medical assistant HealthPal is an example of the first domain. This dialogue based monitoring system aims at supporting elderly people in their preferred environment [2]. Another example is proposed in [3], where the system provides help for younger people suffering from overweight.

P. Kostkova (Ed.): eHealth 2009, LNICST 27, pp. 13–20, 2010.

### 1.2 Transmission of Physiological Parameters

An example of the second domain has been implemented within the MOEBIUS project (Mobile extranet-based integrated user services), which integrates doctors and patients by submitting different physiological parameters [4].

The use of mobile devices in order to assist young cancer patients is described in [5] and the authors conclude that such a system has a number of advantages:

- Higher compliance of appointments with alerting functionality.
- Higher data quality.
- Less work on part of the doctor to prepare the documentation.
- Fewer errors in the documentation.

### 1.3 Alert of Patients or Medical Professionals with Critical Values

An architectural and conceptual overview of an application of the third domain is outlined in [6] and focuses mainly around the actors patient, doctor and nurse. The use cases for a mobile alerting system are built around them.

Another example is SAPHIRE, a monitoring and decision support environment that generates alerts if the patient's state becomes critical in a home-based scenario. It bases on guidelines that are able to evaluate the incoming data and infer critical states. The communication with a clinic able to handle the situation is solved using a secure VPN connection; potential updates to the guidelines are automatically downloaded upon start-up of the application and are therefore always up-to-date [7].

## 2 The eSana Framework

The eSana framework, developed at the University of Fribourg, allows the creation of applications in the second domain. It is illustrated in figure 1 and the approach has been described in more detail in [8] and [9]. The main goals of mobile medical applications using this framework are as follows:

- Location and time-independent communication between doctor and patient.
- Customizable processes and user interfaces per patient.
- Integration of the patient into the disease management and documentation process (patient empowerment, see level-of-care pyramid in [10]).
- Integration of contextual parameters from the environment of the patient (see [11]).
- Integration of new consumer applications receiving physiological parameters.

The eSana framework considers the needs of several actors, e.g. patients, doctors, nurses, health care administrators, by allowing the development of various mobile medical applications. One abstract scenario is the transmission of physiological parameters. Several medical devices transmit their measurements to the mobile device of the patient (e.g. a cell phone) by using Wireless Personal Area Networks such as Bluetooth. The requested information (e.g. weight, glucose, photo of the current meal) is part of a process description which contains a number of forms as user interfaces; it can either be received from a nearby medical device or entered manually.

Both the process definitions and user interface descriptions are stored external to the application as XML artifacts and can be adapted per patient (e.g. by the doctor).

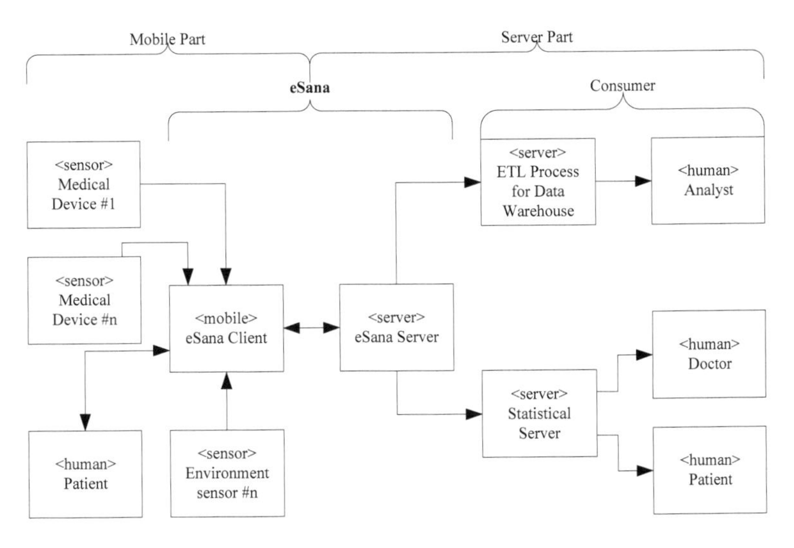


**Fig. 1.** Overview of the eSana framework with example consumers

The process descriptions base conceptually on the UML state event diagram. Each state contains a reference to a user interface described in a separate artifact.

The eSana framework also specifies and interprets a third artifact containing information about the entry screen of the application. It allows the inclusion of additional applications for a specific problem field. Therefore, new domain-specific applications can be embedded into the entry screen. An example of such a specific application is a snippet, which retrieves and visualizes all physiological data of the patient.

Once the patient decides to send the data to the server, it is transformed and dispatched to a number of interested subscribers. These are not part of the eSana framework and can be added depending on specific requirements. For example, a subscriber to dermatological information can offer tools to process the incoming image information in a way that the dermatologist can work with it [12].

Security is a fundamental issue for the exchange of medical data and has already been researched for similar settings (see [13]). It can be summarized by using the following technologies mapped onto the approach above:

- Use of Bluetooth security between the medical and mobile devices.
- HTTPS with client and server X.509 certificates between mobile device and server and also between servers on different systems.
- HTTPS between the service providers and the end users. Whether client X.509 certificates will be used depends on the service offered. Typically, an application communicating with a medical expert will require a stronger authenticity.

## 3 Survey

Based on the possibilities available with the eSana framework, a survey was conducted in Turkey in order to analyze what medical professionals think about using a mobile application in the specific case of diabetes. Other surveys were already conducted in Switzerland (for cardiology [14] and dermatology [15]), but lacked two important aspects: (1) Switzerland has virtually no rural areas with difficult access to healthcare and (2) the health system in Turkey is in a state of paradigm change and open to new ways of integrating ICT in order to improve medical access.

### 3.1 Introduction

The survey was conducted in August and September 2008 in Turkey. In order to get a good response rate, the interviews with the medical professionals were conducted face to face with the help of the ministry of health in Turkey. The sample was chosen from the following three cities near rural areas: Adiyaman (eastern representative), Bolu and Duzce (both western representatives). These cities were chosen because they represent the regions that were transformed first within the new health system in Turkey. All chosen medical professionals are family doctors and are therefore regularly confronted with diabetes cases.

336 doctors were included in the sample. Of these, 239 have participated in the survey resulting in a response rate of 71.1%. The rate does not differ significantly between the cities. 72.4% of the respondents are male and the overall average age is 34 years. In average, they have 8.4 years of medical experience. The vast majority (98.7%) works in a Family Medical Center Unit.

### 3.2 Diabetes Application Context

Several questions were asked about the diabetes use case. The first question was whether and how the respondents were regularly tracking scientific development of diabetes. 77.2% answered that they did. Of these, 70.9% answered that they use publications to keep up to date. 48.0% use electronic databases, 33.5% attend conferences on the subject and 27.9% follow courses in order to extend their knowledge.

When asked, why they think that their patients do not come regularly for the visit, the following answers were given:

**Table 1.** Reasons why diabetics do not follow up regularly (ordered)

| Reason | Percentage |
|---|---|
| The patient is doesn't know the importance of his disease | 72.0% |
| The patient chooses to ignore his disease | 65.7% |
| The patient lives in a rural area | 54.8% |
| There are traditional treatments in place | 38.5% |
| The patient has a low self-control | 16.7% |
| The patient is unable to accept his disease | 15.5% |
| The patient is afraid of his disease | 13.0% |

Regarding patients that live in rural areas, only 27.7% of the doctors think that they are able to make regular visits.

The respondents were also asked about how they think the situation may be improved. 34.3% think that communication must be improved generally. Only about one fifth (20.1%) think it makes sense if doctors are made available at regular intervals in the rural areas. The majority of 61.5% thinks that setting up a solution that allows sending the necessary parameters (e.g. blood sugar level, weight) using mobile technologies, including a feedback mechanism from and to the doctors, makes most sense for patients living in rural areas.

On the final question, whether the respondents would like to trace their patients remotely using ICT such as internet, mobile technologies or videoconferencing, 79.6% answered yes against 20.4% not thinking it makes sense.

### 3.3 ICT Infrastructure and Know-How of the Medical Professionals

In another part of the questionnaire, the respondents were asked whether they know specific technical terms and to rate their knowledge. Not surprisingly, more than 80% of the respondents said that they were using regularly (as both power and normal users) the following technologies: Mobile phones, SMS, Internet, Mail, MSN (chat). Mostly unknown (less than 20% of the respondents answered that they were average or good users) were the terms PDA, GPRS, EDGE. This result is also not surprising, as the terms are quite technical. The remaining terms are also known to a somewhat lesser extent: Bluetooth is used by 60.3%, MMS are known to 52.7%, WAP to 49.5% and the term Browser is only known to 42.3% (probably confusion with the term "Internet") of the respondents.

Only 14.6% think that ICT is used sufficiently in their daily work, 67.4% disagree (the rest is indecisive). However, 98.2% of all respondents have access to personal computers at the place of work, which indicates a rather good setup of the infrastructure. The computers are also heavily used: 58.1% of the respondents use them for more than 4 hours and another 20.9% between 3 and 4 hours every day.

### 3.4 Use of Mobile Technologies

The respondents use their mobile phones regularly. 76.6% answered that they always use it and 23.0% use it at least sometimes. The usage pattern indicates that the most used service is of course the normal voice phone call. Nonetheless, 65.3% use SMS regularly and even Bluetooth is used by 33.9% of all respondents.

When asked about their patients, the respondents said in 68.8% of all cases that their patients use their mobile phones to communicate with them. Of these, 89.6% of the patients normally use voice calls and 9.7% additionally use SMS to communicate with their doctors.

### 3.5 Potential Effects of a Mobile Diabetes Solution

A potential mobile solution to trace diabetes in rural areas was presented and evaluated in a number of questions. In one such question, the respondents were asked whether they think that such an application may reduce the number of deaths from diabetes or one of its symptoms. 29.6% do not think that such an application has any effect on the number of deaths, whereas 8.2% think it definitively does. The majority of 56.2% think that such an application may have a positive effect on reducing deaths.

Whether such an application would improve the relationship between the doctor and patients was asked in the next question. 71.0% think it would improve the relationship, whereas only 5.2% think the effects would be negative. 16.5% do not think that there would be any change in their relationship and 7.4% have no opinion.

When comparing the traditional tracing methods of diabetics with a mobile application in rural areas, 64.9% of the respondents think that the tracing would be facilitated against 10.8% thinking that the tracing would be complicated. 11.3% do not think that it differs much and 13.1% had no opinion.

An important factor for the acceptance of such a system is the question, whether the medical professionals think that such a system would cause more or less work. 74.7% think that their work load will increase, whereas only 19.7% think that it will decrease. Supporting a running environment is a complex issue and does not include considerations such as technical support of the patients.

Furthermore, the respondents were asked about possible effects of using such a mobile application and were able to answer that they agree, disagree or are not sure.

**Table 2.** Various aspects of using a mobile application to trace diabetics (ordered)

| Aspect | Agree | Indecisive | Disagree |
|---|---|---|---|
| Allows the patient to feel more secure | 68.8% | 19.9% | 11.3% |
| Reduces complications of the disease | 65.4% | 24.2% | 10.4% |
| Patient becomes more conscious of his disease | 64.5% | 26.8% | 8.7% |
| Facilitates tracing of patients in rural areas | 62.8% | 25.5% | 11.7% |
| Slows down progress of the disease | 58.4% | 27.3% | 14.3% |
| Morbidity of diabetes will be decreased | 57.0% | 31.7% | 10.9% |
| Mortality of diabetes will be decreased | 55.5% | 33.6% | 10.9% |

In the final question about the effects of such an application, the respondents were asked about the potential improvements for specific aspects of the disease. They were able to give a number like 1 (has no effect), 5 (has partial effect) and 10 (has clear effect). The next table illustrates the median values:

**Table 3.** Possible effects of a mobile application to various disease aspects (ordered)

| Aspect | Median |
|---|---|
| Foot ulcers, slowly recovering wounds | 7 |
| Increase in heart disease, heart attack or apoplexy risk. | 6.5 |
| Retinopathy (Damage to the eye) | 5 |
| Nephropathy (Damage or disease of the kidneys) | 5 |
| Neuropathy (disorders of the nerves of the peripheral nervous system) | 5 |

# 4 Conclusion and Outlook

## 4.1 Interpretation of the Survey Results

The survey results indicate several shortcomings in the current treatment of patients with diabetes. Patients in rural areas have difficulty or little motivation for regular medical visits and their knowledge about their disease is limited or is considered as a minor ailment.

On the other hand, medical professionals appear to have a good understanding of ICT and would therefore be able to instruct their patients to use such a mobile system.

Furthermore, they would be capable of analyzing the data using new technologies. One potential problem lies in the possible increased workload for the doctors. This can be limited by integrating the incoming information into the documentation process. Nonetheless, the current working processes need probably to be adapted in order to allow for some time (e.g. once weekly) to analyze the data of remote patients.

One big problem appears to be the missing awareness of patients for their disease. A prevention campaign can be partly supported by mobile communications. Technically, this could be done using SMS. Furthermore, the eSana framework allows the integration of dedicated application snippets. One such snippet could for example offer the patient a selection of possible meals that are adequate for his condition.

Certain findings need to be analyzed further using more specialized methods, for example why that 29.6% of all medical experts do not think that such an application may reduce the number of deaths or why 20.4% do not wish to track their patients using ICT technologies.

## 4.2 Components of a Mobile Diabetes Application

Based on the results of the survey, this section discusses possible components of a mobile diabetes application for patients in rural areas.

- Process to register the current condition of the patient by measuring his physiological parameters. This process can be personalized per patient. It should include a user interface that allows capturing foot ulcer images using the integrated camera of the mobile device, if applicable for the patient.
- Three additional applications: one that informs the patient about possible meals he can cook in order to improve his medical condition and another that allows the patient to view graphically a simplified view of his physiological parameters. The last application offers a small quiz in order to raise the awareness of his condition.
- A mailbox integrated into the diabetes application that allows a two-way communication between the actors. This would allow the doctor to write a short note to the patient after analyzing his data, if necessary. The communication is secured and may also include voice messages.

The server based application that acts as a consumer of the physiological parameters and is used by the medical professional should include the following functionality in a web interface or as dedicated client application:

- Tool for doctors to analyze the incoming parameters graphically in a browser and to set critical limits per patient with the escalation procedures (e.g. e-mail, SMS).
- Processing possibilities to analyze incoming foot ulcer images and mechanisms to make these images available to dermatologists.
- Mailbox that allows to send messages to patients and to read theirs. It may make sense to integrate this into the messaging solution of the doctor (e.g. MS Outlook).
- Possible export functionality of the parameters to a patient information system. The data is exported in the HL7 standard [12, 16]. Such functionality would greatly improve the documentation of the patient's condition and his medical history.

## 4.3 Outlook

The results of the survey will be analyzed in more detail within a Master thesis at the University of Fribourg in Switzerland. This thesis will also contain a model for such a medical application.

The eSana project is part of an ongoing dissertation and will be further refined. Finally, the LoCa research project, done by the University of Fribourg in collaboration with the University of Basel, analyzes the use of contextual information in a healthcare setting in order to allow actors to use adaptable workflows in their smart home or smart hospital environment.

## References


1. Word Health Organization: eHealth Tools & Services, http://www.who.int/kms/initiatives/ehealth/en (last accessed April 1, 2009)
2. Komninos, A., Stamou, S.: HealthPal: An Intelligent Personal Medical Assistant for Supporting the Self-Monitoring of Healthcare in the Aging Society. In: Proceedings of UbiHealth 2006: The 4th International Workshop on Ubiquitous Computing for Pervasive Healthcare Applications (2006)
3. Königsmann, T., Lindert, F., Walter, R., Kriebel, R.: Hilfe zur Selbsthilfe als Konzept für einen Adipositas-Begleiter. In: HMD – Praxis der Wirtschaftsinformatik, vol. 251, pp. 64–76 (2006)
4. Fischer, H.R., Reichlin, S., Gutzwiller, J.P., Dyson, A., Beglinger, C.: Telemedicine as a new possibility to improve health care delivery. In: M-Health – Emerging Mobile Health Systems, Biomedical Engineering, pp. 203–218. Springer, Heidelberg (2006)
5. Leimeister, J.M., Krcmar, H., Horsch, A., Kuhn, K.: Mobile IT-Systeme im Gesundheitswesen, mobile Systeme für Patienten. HMD – Praxis der- Wirtschaftsinformatik 244, 74–85 (2005)
6. Jung, D., Hinze, A.: A Mobile Alerting System for the Support of Patients with Chronic Conditions. In: Proceedings of the Euro MGOV 2005, pp. 264–274 (2005)
7. Hein, A., Nee, O., Willemsen, D., Scheffold, T., Dogac, A., Laleci, G.B.: SAPHIRE – Intelligent Healthcare Monitoring based on Semantic Interoperability Platform – The Homecare Scenario. In: Proceedings of the ECEH 2006, pp. 191–202 (2006)
8. Savini, M., Ionas, A., Meier, A., Pop, C., Stormer, H.: The eSana Framework: Mobile Services in eHealth using SOA. In: Proceedings of the EURO mGOV 2006 (2006)
9. Stormer, H., Ionas, A., Meier, A.: Mobile Services for a Medical Communication Center – The eSana project. In: Proceedings of the First European Conference on Mobile Government (2005)
10. Rittweger, R., Daugs, A.: Patientenorientiertes Disease Management. In: Jähn, K., Nagel, E. (eds.) e-Health, ch. 3, pp. 162–165. Springer, Heidelberg (2004)
11. Savini, M., Stormer, H., Meier, A.: Integrating Context Information in a Mobile Environment using the eSana Framework. In: Proceedings of the ECEH 2007, pp. 131–142 (2007)
12. Savini, M., Vogt, J., Wenger, D.: Using the eSana framework in Dermatology to improve the Information Flow between Patients and Doctors. In: Proceedings of the Bled eConference, pp. 156–169 (2008)
13. Marti, R., Delgado, J., Perramons, X.: Security Specifications and Implementation for Mobile eHealth Services. In: EEE, vol. 00, pp. 241–248 (2004)
14. Wenger, D., Meier, A., Widmer, M., Savini, M., Dietlin, C.: Einsatz von Informations- und Kommunikationstechnologien (IKT) bei Schweizer Dermatologen. In: Dermatologica Helvetica, p. 8 (2008)
15. von Burg, O., Savini, M., Stormer, H., Meier, A.: Introducing a Mobile System for the Early Detection of Cardiac Disorders as a Precaution from a Cardiologists' View (Evaluation of a Survey). In: Proceedings of the HealthInf2008 – International Conference on Health Informatics (2008)
16. HL7 Inc.: HL7 Messaging Standard Version 2.5.1, an application protocol for Electronic Data Exchange in Healthcare Environments (2007)

# Early Warning and Outbreak Detection Using Social Networking Websites: The Potential of Twitter

Ed de Quincey and Patty Kostkova

City eHealth Research Centre, City University
Northampton Square, London EC1V 0HB
ed.de.quincey@city.ac.uk

**Abstract.** Epidemic Intelligence is being used to gather information about potential diseases outbreaks from both formal and increasingly informal sources. A potential addition to these informal sources are social networking sites such as Facebook and Twitter. In this paper we describe a method for extracting messages, called "tweets" from the Twitter website and the results of a pilot study which collected over 135,000 tweets in a week during the current Swine Flu pandemic.

**Keywords:** Epidemic Intelligence, social networking, swine flu.

## 1 Introduction

Epidemic Intelligence (EI) is being used by public health authorities to gather information regarding disease activity, early warning and infectious disease outbreak [1, 2, 3, 4]. EI systems systematically gather official reports and rumours of suspected outbreaks from a wide range of formal and increasingly, informal sources[1] [5]. Tools such as the Global Public Health Intelligence Network (GPHIN) and Medisys gather data from global media sources such as news wires and web sites to identify information about disease outbreaks [5, 6].

A potential improvement to these systems has been demonstrated by Google's Flu Trends research that has estimated flu activity via aggregating live online search queries for keywords relating to flu [7]. The drawback however is that the information stored in commercial search query logs, which could be integrated into EI systems is not freely available.

The increase in user-generated content on the web via social networking services such as Facebook and Twitter, however provides EI systems with a highly accessible source of real-time online activity. Twitter [8], a micro-blogging service that allows people to post and read other users' 140 character messages, called "tweets", currently has over 15 million unique users per month [9]. Twitter allow third parties to search user messages and return the text along with information about the poster, such as their location, in a format that can be easily stored and analysed.

---

[1] Informal sources account for more than 60% of the initial outbreak reports [5].

P. Kostkova (Ed.): eHealth 2009, LNICST 27, pp. 21–24, 2010.

In this paper we detail the information that is accessible via services such as Twitter, a process that can be used to access it and present the results of a pilot study into identifying trends of flu activity in May 2009, present in messages sent via Twitter.

## 2 Methodology

Twitter allows access to users' tweets via Application Programming Interfaces (APIs): a REST API and the Search API. The REST API method allows developers to "access core Twitter data" [10] such as user profile information, ability to post tweets etc.. The Search API, which is utilised in this paper, allows developers to query tweets in real-time using any combination of keywords. This is via making a request to a url in the following format:

```
http://search.twitter.com/search?q=keyword
```

A number of other parameters can also be passed via the querystring such as number of results to return e.g. `rpp=100`. The matching tweets, containing the text of the tweet, user information and a timestamp, are returned in either atom (an xml format) or json (a computer data interchange format). This data can then be parsed programmatically using PHP, Ruby, C etc..

### 2.1 Use of Twitter in This Study

For this preliminary study, the Search API was utilised to return the last one hundred tweets that contained instances of the word "flu". PHP code was then written to parse the returned tweets (in atom format) and save them to a MYSQL database, comprised of one table. Records collected comprised of the following fields:

```
id, published, link, title, content, author, terms
```

A batch file was created that ran the PHP code every minute with new tweets being saved[2] in the database. The program was started at 14:00 on Thursday 7th May 2009 and has been running continuously since then. The results presented in this paper are taken from the following week, i.e. until 14:00 on Thursday 14th May 2009.

## 3 Preliminary Results

During the week, there were a total of 135,438 tweets, posted by 70,756 unique users that contained the word "flu" with the following table showing their daily distribution.

The lowest number of tweets was recorded on Sunday the 10th of May (discounting the 14th as only 14 hours of tweets was collected) and the highest on Friday the 8th of May.

---

[2] It was found that this rate was sufficient as it was unlikely that there were more than one hundred new tweets in a minute.

**Table 1.** Number of tweets containing the word "flu

| Date | Number of Tweets |
|---|---|
| Thursday 7th May 2009[3] | 16,422 |
| Friday 8th May 2009 | 24,692 |
| Saturday 9th May 2009 | 18,484 |
| Sunday 10th May 2009 | 15,213 |
| Monday 11th May 2009 | 19,140 |
| Tuesday 12th May 2009 | 19,353 |
| Wednesday 13th May 2009 | 14,370 |
| Thursday 14th May 2009[3] | 7,764 |
| **Total** | 135,438 |

The use of the word "flu" however varied greatly in the tweets with users utilising the term to refer to themselves, a friend, a news story, a link etc.. Further analysis into the actual meaning of the tweets is currently being planned but to identify any immediate trends, the content of all the tweets was analysed using concordance software. The following table shows a selection of the top words present in all of the tweets (common words such as "a", "the", "to" etc. have been removed).

**Table 2.** Most popular words found in all tweets

| Word | Frequency | Word | Frequency |
|---|---|---|---|
| Flu | 138,260 | New | 7,668 |
| Swine | 99,179 | News | 6,498 |
| Have | 13,534 | Confirmed | 6,456 |
| Cases | 13,300 | Just | 6,373 |
| H1N1 | 9,134 | People | 5820 |
| Has | 8,010 | Case | 5647 |

In the majority of tweets the word "swine" was present along with "flu" (which would perhaps be expected with the current swine flu pandemic). Although the word "have" is considered to be a common word in the English language (24th most common [11]), it has been included in this list because it might be an indication of people tweeting that they "have flu"[4]. For a similar reason the use of the word "has" may indicate that the tweet contains information about someone else having flu e.g. "he has flu". The words "confirmed" and "case(s)" perhaps indicate a number of tweets that are publicising "confirmed cases of swine flu". Further investigation into this is being conducted using collocation analysis, a sample of which is shown in the following table (again excluding common words).

---

[3] Recording of tweets began at 14:00 on 7/5/2009 and stopped at 14:00 14/5/2009.
[4] Interestingly the phrase "have flu" was found only 137 times.

**Table 3.** Collocation of words, one word to the right and the left of the word "flu"

| 1 word to the left | | 1 word to the right | |
|---|---|---|---|
| **Word** | **Frequency** | **Word** | **Frequency** |
| Swine | 96,651 | Http | 6,598 |
| The | 5,701 | Cases | 6,194 |
| H1n1 | 5,225 | Case | 2,210 |
| Bird | 1,425 | Death | 2,001 |
| New | 1,304 | Virus | 1,411 |
| Pig | 1,164 | Outbreak | 1,321 |
| Man | 720 | H1n1 | 1,147 |
| Stomach | 510 | Spreads | 927 |
| Regular | 426 | Lol | 924 |
| Flu-bird | 319 | Deaths | 912 |

## 4 Conclusion

The results described in the previous section highlight the potential for twitter to be used in conjunction with pre-existing EI tools. Although a potential explanation of the number of tweets collected could be due to the current swine flu pandemic, the amount of real-time information present on twitter, either with regards to users reporting their own illness, the illness of others or reporting confirmed cases from the media, is both rich and highly accessible. Further work is planned into the data already collected and the system is continually retrieving and storing tweets to be analysed in relation to users' geographical location and the semantic syntax of tweets.

## References


1. Linge, J.P., Steinberger, R., Weber, T.P., Yangarber, R., van der Goot, E., Al Khudhairy, D.H., Stilianakis, N.: Internet surveillance systems for early alerting of health threats. Euro Surveill. 14(13), pii=19162 (2009)
2. Kaiser, R., Coulombier, D.: Different approaches to gathering epidemic intelligence in Europe. Euro Surveill. 11(17), pii=2948 (2006)
3. Paquet, C., Coulombier, D., Kaiser, R., Ciotti, M.: Epidemic intelligence: a new framework for strengthening disease surveillance in Europe. Euro Surveill. 11(12), pii=665 (2006)
4. Coulombier, D., Pinto, A., Valenciano, M.: Epidemiological surveillance during humanitarian emergencies. Médecine tropicale: revue du Corps de santé colonial 62(4), 391–395 (2002)
5. WHO, `http://www.who.int/csr/alertresponse/epidemicintelligence/en/index.html`
6. Linge, J.P., Steinberger, R., Weber, T.P., Yangarber, R., van der Goot, E., Al Khudhairy, D.H., Stilianakis, N.: Internet surveillance systems for early alerting of health threats. Euro Surveill. 14(13), pii=1916 (2009)
7. Google Flu Trends, `http://www.google.org/flutrends/`
8. Twitter, `http://www.twitter.com`
9. `http://www.crunchbase.com/company/twitter`
10. Williams, D.: API Overview, `http://apiwiki.twitter.com/API-Overview`
11. `http://www.world-english.org/english500.htm`

# Communicating with Public Health Organizations: An Inventory of Capacities in the European Countries

Daniel Catalan Matamoros, Alexandru Mihai, Skaidra Kurapkiene, and Wadih Felfly

European Centre for Disease Prevention and Control, Director's Cabinet,
Tomtebodavagen 11A, SE-171 83, Stockholm, Sweden
{daniel.catalan,Alexandru.mihai,skaidra.kurapkiene,
wadih.felfly}@ecdc.europa.eu

**Abstract.** The European Centre of Disease Prevention and Control (ECDC) is an EU agency established in 2005 with the mission to identify, assess and communicate current and emerging threats to human health posed by infection diseases. In order to keep high quality relations with our stakeholders, it was identified the crucial need of creating a database with relevant, updated and reliable information about the contacts and organizations in the Member States. The ECDC Contacts and Organizations Database (ECO DB) aims to avoid duplications, overlapping and to increase the quality in the communication with the ECDC's stakeholders. Microsoft CRM is the IT platform used to create the ECO DB. CRM provides a competitive set of functionalities which make this IT tool an excellent solution for contacts management which facilitates and improves the communication between the Member States and the ECDC.

**Keywords:** Information architecture, Public Health, communication.

## 1 Introduction

The European Centre of Disease Prevention and Control (ECDC) is an EU agency established in 2005 with the mission to identify, assess and communicate current and emerging threats to human health posed by infection diseases. In order to achieve this mission, ECDC works in partnership with national protection bodies across Europe to strengthen and develop continent-wide disease surveillance and early warning systems. By working with experts throughout Europe, ECDC pools Europe's health knowledge, so as to develop authoritative scientific opinions about the risks posed by current and emerging infectious diseases.

ECDC activities rely on coordination and constantly involve communication and exchange of information with many external organizations and people. ECDC is a fast growing organization, both in terms of people and functional coverage and activities.

The conditions for collecting, storing and maintaining of information are rapidly changing. It is vital to choose the proper information technology tools for collecting and distributing constantly growing information, since the stored and maintained data must be true, reliable and updated. Accurate and timely information provided in a professional manner is the key to any service operation.

P. Kostkova (Ed.): eHealth 2009, LNICST 27, pp. 25–27, 2010.

Customer Relationship Management (CRM) software might be the right solution [1]. CRM highlights the importance of using information technologies in creating, maintaining and enhancing customer relationships. However, there is a need to develop a better understanding of CRM and how public health companies can use IT tools such as CRM.

There are many ways to use CRM. The most important aspect to understand is how to get the most out of the database and to accept CRM as a business philosophy, rather than a marketing strategy.

## 2 The Use of Customer Relationship Management in Public Health

There are some other experiences in similar environments showing multiple ways of using and functioning of CRM database for the organizations' information management. For example, CRM can be implemented in non-profit organizations such as primary health care centers and hospitals [2]. Health care organizations use CRM to research and analyze capital investment decisions in ways that are not possible using only an internal operating system, to enhance direct mail, to create a strong physician-patient relationship, to help its organization retain valuable patients. Other health providers improved the interactivity of their Web site using CRM to meet the needs of their market [3]. A Midwestern U.S. hospital used its CRM database marketing program to identify those with the greatest likelihood to use or need cardiology services, coupling direct mail and an online Health Risk Assessment [4]. CRM enables the health care industry to get essential customer information and use it as efficiently as possible, thus it enables the health care sector to improve patient health, increase patient loyalty and patient retention and add new services as well [5].

## 3 The ECDC Contacts and Organizations Database

The European Centre for Disease Prevention and Control (ECDC) started operations in May 2005 and has since then gone through a quick start-up phase building its organization infrastructure, developing its operational principles, tools and procedures, hiring the core staff of experts and creating a solid programme framework in surveillance, scientific advice, emergency preparedness, response and training.

For its daily activities ECDC requests high quality work in the field of country relations and information. In order to reach this goal, it has been identified the crucial need of creating a database with relevant, constantly updated and reliable country information consisting of the contacts, organizations and other appropriate information in the Member States. The ECDC Country information system is the consequence of the decision that all the ECDC contacts and organizations have to be on a common database in order to avoid duplications, overlapping and to increase the quality in the communication with the ECDC's stakeholders. The objective is to build an ECDC-wide public health country inventory that provides high quality and easy to access information on all relevant aspects for ECDC.

Two years ago, ECDC decided to adopt Microsoft CRM as the IT product used to create the country information database. CRM is a term applied to processes implemented by a company to handle its contact with its customers. In ECDC, the CRM software is used to support these processes and to store information on partners. All information in the system is both accessed and entered by staff from the different units. It is easy to implement, and to connect to other products used in ECDC, such as Word, Excel, SharePoint and any SQL database.

CRM provides a competitive set of functionalities: tracking the email exchange with all ECDC contacts; contacts ownership and update processes; integration with other in-house systems (Identity Management, The European Surveillance System, Epidemic Intelligence Information System, Threat Tracking Tool, Web portal, Intranet, Terminology server, Document management system), etc. These functionalities make CRM an excellent tool for contacts management which improves the communication between the Member States and the ECDC.

An operational database of country contacts, resources and capacities based on CRM includes the developing and maintaining an inventory on communicable diseases in member states.

The main data sets included are: Contacts and organizations, reference laboratories, experts' directory, medical libraries, national media, and some other relevant information sets in the field of public health capacities.

The advantages of the Country Information in CRM are:

- Centralization of system administration;
- Easy monitoring of availability and capacity;
- Easily customizable to the ECDC needs;
- Quick response time.

As a result, this environment facilitated by CRM is increasing the added value of the work of country relations in public health.

## References


1. Customer Relationship Management: how a CRM system can be used in the sales process, http://www.essays.se/essay/f3d2ddb07f/ (accessed February 1, 2009)
2. Sopacua, E.: The Implementation of CRM in Public Health Centers and Hospitals as an Alternative for Service Marketing Strategy, http://www.journal.unair.ac.id/detail_jurnal.php?id=2240&med=6&bid=3 (accessed April 2, 2009)
3. Atwood, A.: InfoManagement Direct, Benefits of an Innovative CRM Solution for Health Care Organizations, http://www.information-management.com/infodirect/20061103/1066948-1.html (accessed February 1, 2009)
4. CRM software solutions, http://www.databasesystemscorp.com/tech_crm_applications_219.htm
5. Health Care Industry Opts for CRM! http://www.crminfoline.com/crm-articles/crm-health-care.htm (accessed April 2, 2009)

# Transparency and Documentation in Simulations of Infectious Disease Outbreaks: Towards Evidence-Based Public Health Decisions and Communications

Joakim Ekberg[1], Toomas Timpka[1], Magnus Morin[3], Johan Jenvald[3], James M. Nyce[4], Elin A. Gursky[5], and Henrik Eriksson[2]

[1] Dept. of Medicine and Health Sciences, Linköping University, SE-581 83 Linköping, Sweden
joakim.ekberg@liu.se, tti@ida.liu.se
[2] Dept. of Computer and Information Science, Linköping University, Linköping, Sweden
her@ida.liu.se
[3] VSL Research Labs, Linköping, Sweden
magnus.morin@vsl.se, johan.jenvald@vsl.se
[4] Dept. of Anthropology, Ball State University, Muncie, IN, USA
jnyce@rocketmail.com
[5] National Strategies Support Directorate, ANSER/Analytic Services Inc, Arlington, VA, USA
elin.gursky@anser.org

**Abstract.** Computer simulations have emerged as important tools in the preparation for outbreaks of infectious disease. To support the collaborative planning and responding to the outbreaks, reports from simulations need to be transparent (accessible) with regard to the underlying parametric settings. This paper presents a design for generation of simulation reports where the background settings used in the simulation models are automatically visualized. We extended the ontology-management system Protégé to tag different settings into categories, and included these in report generation in parallel to the simulation outcomes. The report generator takes advantage of an XSLT specification and collects the documentation of the particular simulation settings into abridged XMLs including also summarized results. We conclude that even though inclusion of critical background settings in reports may not increase the accuracy of infectious disease simulations, it can prevent misunderstandings and less than optimal public health decisions.

**Keywords:** outbreak simulation, ontologies, report generator.

## 1 Introduction

Recognizing the threat of a destructive pandemic influenza outbreak, the World Health Organization has urged countries to develop preparedness plans [1]. In these preparations, computer simulations have emerged as an important tool for analyzing competing interventions, producing forecasts to be used in exercises, and for support of policy making [2].

Simulations of outbreaks of infectious disease can be powerful tools in local and regional pandemic-response planning [3], but to be most useful for this purpose, there

P. Kostkova (Ed.): eHealth 2009, LNICST 27, pp. 28–34, 2010.

must be methods for disseminating the simulation results. Without a sufficiently transparent documentation procedure, the advantages simulations offer public health planners become less than desired. In particular, simulation reports must be truthful and clear with regard to the simulation parameters and their values, even when the reports are detached from the immediate simulation context.

The research question this study addresses is how simulation results can be reported responsibly to multiple target audiences, such as when the reports are used as decision support in policy-making contexts. The specific aim is to investigate how the settings used in simulation models can be incorporated in the generation of simulation reports. Previously, we reported an ontology-based approach to outbreak simulation with separate models [4,5] specifically designed to support exploration of alternative interventions under varying hypothetical conditions in local communities [6,7]. This paper describes the method used for results dissemination, in particular a procedure for routine compilation and summary of simulation results.

## 2 Methods

We extended the existing simulator architecture [5] with a prototypical simulation report generator. The architecture uses the ontology management system Protégé [8] for representing and manipulating simulator settings. The simulator settings are transferred to a simulation engine, designed for computational efficiency, which runs the simulation according to the settings. As a step in an overall assumptions management effort [9], the implemented report generator can access both settings and results to produce a report from the simulation displaying specified subset(s) of the underlying assumptions.

In the development of the report generator, we took into account that many different sets of factors can be chosen to be modelled. There are many factors that theoretically may affect the progress of an actual infectious disease outbreak. The problem is not only to enumerate relevant factors, but also to select significant (i.e., sufficiently important) factors, and to make comparisons possible between different results. In our ontology-based approach to simulation, we separate models for the community, the disease, and interventions. The community model can be made to match factual communities to varying degrees. The disease model includes the epidemiological data available for the specific disease modelled and the intervention model contains different strategies modelled and available and accessible to policy makers.

The Models of Infectious Disease Agent Study (MIDAS) project has documented model profiles, which are standardized descriptions of simulation models devised to aid comparisons between models by detailing modelling, assumptions, data sources and implementation issues [https://www.epimodels.org/midas/modelProfilesFull.do]. We separated components from these profiles and used these to develop a typology of explicit simulator settings [9].

The ontology management system of the simulator was augmented by devising a Protégé extension for tagging different categories of settings. The simulator produces documentation from each step in the simulation process in a standardized format using XML (eXtensible Markup Language) documents, thus making a range of trace information available for post-processing and inspection.

For implementing the prototypical report generator, we used the scripting language XSLT (eXtensible Stylesheet Language Transformations), which has been specifically designed for transforming XML documents. The task of compiling simulation results and settings to produce simulation reports in various forms was carried out by a collection of XSLT scripts.

# 3 Results

The augmented simulation environment keeps track of associations between settings and simulation results. For instance, it is recognized (and shows policy makers) that the reporting of transmission rates is dependent on settings in the simulation model regarding sources for population data, transmission probabilities between different age groups, behavioural settings, such as proportion of individuals staying home from work, and intervention settings.

## 3.1 Settings and User Groups

For functional reasons, the parametric settings governing simulations of infectious disease transmission were divided into broad categories that reflect the response planning situation. These settings reflect a condition that can be made subject for intervention (is modifiable) and if the setting is based on verified empirical observations or heuristically estimated.

*Model settings* address the level at which communities, diseases, and interventions need to be represented. The settings include the selected granularity of population data and mixing group structure. For instance, it is often assumed that schools constitute a homogenous mixing group with regard to infectious disease transmission [10]. The community model settings are often based on reports of the daily close contact probabilities for pairs of individuals from different age groups [6]. But the inclusion of too much detail means that simulations not only become more complex, but also can impede validation against empirical data. The model settings can be varied in the simulation environment through substitutable components of community models or disease models. Model assumptions and paradigm assumptions are usually fundamentally associated with the simulation context to be interpreted and analyzed in most policy-making contexts. Comparisons of analyses using different community model settings are also possible but mostly for research and development purposes. Complete reports of settings, with the addition of random seed, software versions, and other technical information may serve as a technical documentation.

*Scenario settings* include fixed environmental assumptions that normally cannot be controlled by the health response managers, for example the surge capacity of the health care system. Using these scenario assumptions, it is possible to evaluate the effect of an intervention by varying the degree of response and efficiency of the response organizations. Scenario assumptions also include non-controllable epidemiological settings based on probable values on biological parameters derived from the current understanding of the infectious disease in question, such as transmissions rates, the average incubation period, and efficacy of pharmaceuticals. Modifiable scenario settings, such as the choice and timing of interventions are the most important model components to be adjusted and compared in reports, because most health policy analyses are designed to be able to explore alternative interventions.

*Behavioural assumptions* address action patterns in populations. These include changes in behaviour due to increased risk awareness during an ongoing pandemic, including voluntary social distancing.

### 3.2 Implementation

In the extended version of the simulation environment, the user defines a simulation job by specifying the simulation model, instance data, and experiment parameters, and submits it to the computational environment, which is a discrete-event simulator designed for efficiency.

The simulation job generates a number of stochastically-generated populations (usually $n$=100) with randomized initial cases. These are run in the computational environment, simulating each specification 1,000 times. All this documentation is collected into a repository of XML documents.

The report generator uses XSLT specification to collect the documentation for the particular simulation settings, including the settings used for the simulation job, the generated mixing group specification for the simulator, and the simulator results of all collections of simulations into a raw XML report. This intermediate raw XML report is thereafter processed to produce different types of abridged XMLs that summarize results and settings. These report XMLs can easily be manipulated and transformed into a range of formats with built-in features of XSLT as illustrated in Figure 2 (e.g., HTML and PDF). In the current approach, the transformation of settings to textual reports does not take advantage of any natural-language processing, but are reported as stated in the simulator user interface.

In the implementation of the automated report generator in XSLT, the stochastically generated population model documentation was about 20 megabytes in size. With 100 generated populations, this size was too large to be practically manageable. To overcome this, we assumed that the simulation results, including information on how age and mixing group members were infected, were sufficient for documentation purposes. Reports of the variation between different generated populations were deemed necessary only when tracing when and how specific individuals transmitted infection.

### 3.3 'Swine flu' Use Example

Involvement of outbreak response specialists in provision of model settings and interpretation of simulation reports can be seen in an analysis of high schools closures in preparation of response to the 'swine flu' outbreak in 2009 (Figure 1). The simulation specialist provides (a) community model settings, the health care manager provides (b) non-modifiable scenario settings, the chief epidemiologist provides (c) non-modifiable epidemiological settings, and the policy-maker specifies (d) the intervention. Figure 2 shows the report presenting the analysis results.

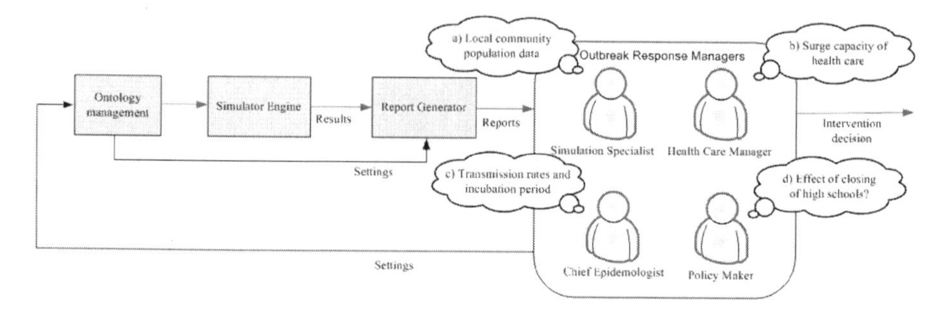


**Fig. 1.** Overview of the 'transparent' predictive modelling process

Simulation report generated from SimulationJob_1/Scenario_1#1 using one SimCore indata XML and all SimCore outdata XML, generated in approx 4 sec using Saxon6.5.5

**Results:**

**Type of simulation: R0**

R0: 1.9965800000000014

Distirbution of mean R0
1.42 1.48 1.57 1.60 1.65 1.66 1.68 1.69 1.72 1.72 1.72 1.73 1.75 1.76 1.78 1.78 1.79 1.79 1.80 1.81
1.81 1.82 1.83 1.84 1.84 1.84 1.84 1.85 1.85 1.85 1.85 1.85 1.87 1.87 1.89 1.89 1.90 1.90 1.91 1.91
1.92 1.93 1.94 1.94 1.94 1.95 1.95 1.96 1.96 1.96 1.97 1.97 2.00 2.00 2.01 2.02 2.02 2.03 2.03 2.03
2.03 2.03 2.04 2.04 2.05 2.05 2.06 2.06 2.07 2.08 2.09 2.10 2.12 2.12 2.12 2.13 2.14 2.14 2.16 2.17
2.17 2.18 2.19 2.19 2.20 2.22 2.25 2.26 2.28 2.33 2.33 2.33 2.33 2.38 2.38 2.44 2.48 2.50 2.59 2.75

Max: 2.757
Min: 1.427
Range: 1.33
Standard diversion between mean R0: 0.2347481427084153
Variance: 0.05510669050505051
Number of generated populations (iterations): 100
Number of simulations for each population (repetitions): 1000
Number of initial cases: 1
Simulation mode (R0 or duration): R0

**Model Settings**

**Community Model**

Population data source: Linköping:no workgroup
Types of Mixing Groups used: Household, Neighborhood, Community, PlayGroup, Daycare, ElementarySchool, MiddleSchool, Highschool

**Epidemiological settings**

base-model H1N1-influenza
Age Groups: [- 5 ] [ 6 - 18 ] [ 19 - 64 ] [ 65 - ]

Types of Mixing Groups used: Household, Neighborhood, Community, PlayGroup, Daycare, ElementarySchool, MiddleSchool, Highschool

**Transmission Rates**

Household

| Age Groups | - 5 | 6 - 18 | 19 - 64 | 65 - |
|---|---|---|---|---|
| - 5 | 0.08 | 0.08 | 0.03 | 0.03 |
| 6 - 18 | 0.08 | 0.08 | 0.03 | 0.03 |
| 19 - 64 | 0.03 | 0.03 | 0.04 | 0.04 |
| 65 - | 0.03 | 0.03 | 0.04 | 0.04 |

Neighborhood

| Age Groups | - 5 | 6 - 18 | 19 - 64 | 65 - |
|---|---|---|---|---|

**Fig. 2.** Example of 'swine flu' analysis report with explicit simulator settings

## 4 Discussion

During an ongoing crisis, such as the 'swine flu' outbreak in 2009, health-care policy-makers at a local level need to deal with complex problems. Typically, they need to get a clear picture of what part of the situation (and a simulation) can be affected by their decisions and what is largely out of their control. Simulations can present the dynamics of pandemic outbreaks for policy makers and help them choose between and prioritize decision alternatives.

To facilitate this, heuristic estimates need to be easily recognizable as factual data becomes available and scenarios can be updated. For instance, among avian influenza cases in Indonesia, the mean incubation period appears to have been approximately 5 days, which is nearly twice as long as for past pandemic strains and current inter-pandemic strains of influenza [11]. Redefinition of the non-modifiable epidemiological settings according to new data can dramatically change the simulation outcome. Without proper documentation of what knowledge source has been used, it becomes impossible to determine on what basis predictive modelling is performed. For instance, during the presently (July 2009) ongoing 'swine flu', reports on reproduction ratio (average number of secondary cases per primary case) and generation intervals (the time between primary and secondary case infection) [12,13] can easily be compared to previously run simulations when new data emerge.

Similarly, the behavioural assumptions made in pandemic simulations tend to overlook the variety of possible behavioural responses to an epidemic and thereby be flawed [14,15]. The result is that these simulations rest on a more or less simplistic representation of human and social behavioural response. Without proper communication of the behavioural assumptions made, simulations may be interpreted as accurate regarding social dynamics when they are not.

The range of possible explicit settings in reports of an infectious-disease simulation precludes any serious attempt to find a complete or even optimal set of explicit simulator settings. Some more or less well founded settings are explicitly modelled, while others by necessity have to be assumed non-variable or non-relevant. Correspondingly, it is possible to categorize the settings used in predictive modelling of infectious diseases in several ways, and the present categorization is not attempted to be theoretically complete. The visualization of possible settings has to be dynamic and accessible, therefore it has to be organized in a both systematic and manageable way. One of the benefits of using an ontology approach is that it is possible to reorganize and inspect the way the settings are categorized and defined.

## 5 Conclusions

Simulation can be a powerful resource that can help decision makers prepare for outbreaks of infectious disease. We have described how the background settings used in simulation models can be incorporated in the generation of simulation reports. This transparency does not necessarily increase the accuracy or validity of simulations per se. However, without transparent and accessible documentation procedures, it is difficult for policy makers to interpret the results. To have this functionality built into simulation tools will improve collaborative planning and the kinds of choices health care decision maker make in outbreaks of infectious disease.

**Acknowledgements.** This work was supported by the Swedish Research Council under contracts 2006-4433 and 2008-5252.

## References

1. WHO. Pandemic influenza preparedness and response WHO guidance document Geneva: World Health Organization (2009)
2. Straetemans, M., Buchholz, U., Reiter, S., Haas, W., Krause, G.: Prioritization strategies for pandemic influenza vaccine in 27 countries of the European Union and the Global Health Security Action Group: a review. BMC Public Health 7(1), 236 (2007)
3. Jenvald, J., Morin, M., Timpka, T., Eriksson, H.: Simulation as decision support in pandemic influenza preparedness and response. In: ISCRAM, Delft, The Netherlands, May 13-16 (2007)
4. Timpka, T., Morin, M., Jenvald, J., Eriksson, H., Gursky, E.: Towards a simulation environment for modeling of local influenza outbreaks. In: AMIA Annu. Symp. Proc., pp. 729–733 (2005)
5. Eriksson, H., Morin, M., Jenvald, J., Gursky, E., Holm, E., Timpka, T.: Ontology based modeling of pandemic simulation scenarios. Stud. Health Technol. Inform. 129, 755–759 (2007)
6. Holm, E., Timpka, T.: A discrete time-space geography for epidemiology: from mixing groups to pockets of local order in pandemic simulations. Stud. Health Technol. Inform. 129, 464–468 (2007)
7. Timpka, T., Morin, M., Jenvald, J., Gursky, E., Eriksson, H.: Dealing with ecological fallacy in preparations for influenza pandemics: use of a flexible environment for adaptation of simulations to household structures in local contexts. Stud. Health Technol. Inform. 129, 218–222 (2007)
8. Gennari, J.H., Musen, M.A., Fergerson, R.W., Grosso, W.E., Crubézy, M., Eriksson, H., Noy, N.F., Tu, S.: The evolution of Protégé: An environment for knowledge-based systems development. Int. J. Hum. Comp. Stud. 58(1), 89–123 (2003)
9. Eriksson, H., Morin, M., Ekberg, J., Jenvald, J., Timpka, T.: Assumptions management in simulation of infectious disease outbreaks. In: AMIA Annual Symp. Proc. (in press, 2009)
10. Germann, T.C., Kadau, K., Longini Jr., I.M., Macken, C.A.: Mitigation strategies for pandemic influenza in the United States. Proc. Natl. Acad. Sci. U S A 103(15), 5935–5940 (2006)
11. Yang, Y., Elizabeth Halloran, M., Sugimoto, J.D., Longini Jr., I.M.: Detecting Human-to-Human Transmission of Avian Influenza A (H5N1). Emerg. Infect. Dis. 13(9), 1348–1353 (2007)
12. Boelle, P., Bernillon, P., Desenclos, J.: A preliminary estimation of the reproduction ratio for new influenza A (H1N1) from the outbreak in Mexico. Euro. Surveill 14(19) (March-April 2009)
13. Fraser, C., Donnelly, C.A., Cauchemez, S., Hanage, W.P., Van Kerkhove, M.D., Hollingsworth, T.D., Griffin, J., Baggaley, R.F., Jenkins, H.E., Lyons, E.J., Jombart, T., Hinsley, W.R., Grassly, N.C., Balloux, F., Ghani, A.C., Ferguson, N.M., Rambaut, A., Pybus, O.G., Lopez-Gatell, H., Apluche-Aranda, C.M., Chapela, I.B., Zavala, E.P., Guevara, D.M., Checchi, F., Garcia, E., Hugonnet, S., Roth, C.: Pandemic Potential of a Strain of Influenza A (H1N1): Early Findings. Science (2009)
14. Ferguson, N.: Capturing human behavior. Nature 446, 733 (2007)
15. Timpka, T., Eriksson, H., Gursky, E., Nyce, J., Morin, M., Jenvald, J., Strömgren, M., Holm, E., Ekberg, J.: Population-based simulations of influenza pandemics: validity and significance for public health policy. Bull. World Health Organ. 87, 305–311 (2009)

# Prototyping a Personal Health Record Taking Social and Usability Perspectives into Account

Enrico Maria Piras, Barbara Purin, Marco Stenico, and Stefano Forti

Fondazione Bruno Kessler (FBK) – 38100 Povo (Trento), Italy
{piras,purin,stenico,forti}@fbk.eu

**Abstract.** This paper presents the process of design involved in prototyping a Personal Health Record (PHR), a patient-centered information and communication hub. As the PHR has to be used by laypeople, we focused on their health related activities (i.e. information management) carried out in the household using a sociological perspective to elicit the infrastructural requirements of the IT. We identified three distinct document management strategies (zero effort, erratic, networking) and 'translated' them into three design characteristics: flexibility, adaptability and customizability. We argue that the key to such PHR success is its capability to support the existing activities carried out by laypeople in managing their health record.

**Keywords:** Personal Health Record, SOA.

## 1 Introduction

Healthcare sectors in western countries are facing a worsening shortage of personnel and constant growing costs. A rising demand for cure, associated with an aging population, requires new organizational solutions to maximize the efficiency of the whole system. These problems are faced also through the adoption of ICTs so to reduce medical errors and increase inter-organizational efficiency and inter-organizational coordination [1]. Still, so far, technologies such as the Electronic Medical Records (EMRs) have failed in creating an integrated healthcare provision as rarely these systems communicate [2].

While research on EMRs' is still a major issue in medical informatics, in the last years there is a growing interest for technologies that address the healthcare sectors' problems proposing to empower patients. Personal Health Record (PHR) is among these. PHRs are electronic patient-controlled hub of information that should allow the individual (or a family member) to access, manage and share their health information [3]. An ideal PHR should support the individuals in their health-related activities through all their lives, adapting to their evolving needs. At the moment there are very few fully functioning PHR system but there is a growing production of different prototypes. Still, even a superficial analyses of the ongoing debate on PHRs shows that there's a strong optimism that relies on the foreseen enthusiasm of laypeople to use this technology. It is clear that this foreseen willingness to adopt a PHR is part of the rhetoric of innovation. To our knowledge, though, few PHRs have been designed after an analysis of what people actually do to manage their own health [4].

P. Kostkova (Ed.): eHealth 2009, LNICST 27, pp. 35–42, 2010.

We assume here that just like EMRs are designed having in mind the workflow and the needs of the medical personnel, PHRs need to take into account the multifarious ways and settings in which the individuals are more likely to use them. At the same time, these systems need to be easily usable by anyone. In this paper we'll present a prototype of TreC, a PHR sponsored by the Autonomous Province of Trento, a local government authority in northeastern Italy. TreC has been designed according to the principles stated above. Data model and architectural components have been chosen after a sociological inquiry about the strategies adopted by laypeople to manage their health information in the household and the analysis of the literature on usability.

## 2 Personal Health Management in the Household

The project started with a research aimed at eliciting the real activities carried out by people to take care of themselves and their perceived hierarchy of relevance. We focused our attention on the health information management, the 'invisible work' [5] people are supposed to do in order to communicate with their doctors, with the purpose to identify the strategies commonly adopted.

### 2.1 Methodology

We analyzed health information management of 30 families conducting ethnographic interviews [6] in their houses. The context of the interaction allowed to combine in-depth interviewing with ethnographic observation. The respondents agreed to show us their medical archives and the places of the houses were they were kept. Interviews (50-80 minutes) were audio recorded and transcribed. Transcripts were coded using a grounded theory method [7] and analyzed using Atlas T.I 5.5 software.

### 2.2 Three Strategies of Health Information Management in the Household

We analyzed the interviews in search of common elements people use to justify the ways they keep medical archives. Four concepts emerged: existence of a classification system, foreseen use of information, perceived importance of the information, and network of caregivers (formal or informal) which would access the documents.

'Classification system' refers both to the actual existence of a way to sort documents out and to its mutability in time. 'Use of information' is the frequency of access to archives, to share information with their doctors or for personal interest. The "perceived relevance", as opposed to the actual use, is the foreseen use of the information. Finally, the "network of caregivers (formal and informal)" refers to the people with whom the documents in the archives are shared. The combination of the four dimensions led us to identify three strategies in health information management.

A "zero effort strategy" is the most adopted. Health information management is simply a record keeping activity without any perceived usefulness of the records kept. Documents are not classified but simply kept randomly (e.g, all in a drawer), people do not browse them in search of information nor their doctors ask for them once again, and people are not able to tell if they are going to be needed. As they are not considered relevant, people do not consider a priority to make them easily accessible.

**Table 1.** Three personal health information management strategies

| | Zero-effort strategy | Erratic strategy | Networking strategy |
|---|---|---|---|
| Classification system | Random and long lasting | Analytical and mutable | Analytical and long lasting |
| Use of information | Rare/null | Frequent | Medium |
| Perceived importance | Low/null | High | High |
| Network of caregivers | One doctor | More than one professional caregiver | A network of informal caregivers |

An “erratic strategy” is characterized by an analytical classification developed to keep the interaction with the doctors running smoothly. People and doctors both consider highly relevant the information stored. Documents have to be re-ordered frequently to use them in the relationships with more than one doctor as each of them is interested only in a sub-set of data. Moreover, the evolution of the disease may suggest a different classification to better suit the need of the moment. The constant use of these data requires them to be easily accessible at any time.

“Networking strategy” is adopted as a part of a shared support for elderly or disabled people. Documents are sorted out in order to provide the relevant information in emergency. Moreover, health records are placed so to be accessible to a given number of informal caregivers (e.g., sons and daughters of an old parent) so that any of them can gather information (e.g. to manage a medication scheme).

These strategies are not mutually exclusive. Rather, many people adopt more than one at time for different records. A cancer patient, for instance, would use an ‘erratic strategy’ for this disease while keeping a ‘zero effort strategy’ to manage other, less relevant, health records.

### 2.3 Implications for System Design

As people may use more than one strategy for health information management the system has to be flexible, allowing users to decide which information is relevant and how to create connections between documents. While a ‘zero effort’ strategy merely requires the documents to be stored, an ‘erratic’ strategy can only be supported by making the system flexible enough to be adapted to the unpredictable changes in one’s health. Moreover, people have to be given the opportunity to customize their system adapting it to the needs of the moment just as they do with their paper records. Finally, the observation conducted while interviewing allowed to see that the relevance of some health document was stressed by strategically placing it in different places in the house. People keep prescription by the entrance door of the house became to remember to buy medications; they keep calendars with (also) medical appointments in the corridor to coordinate the social activities of the family; they keep their diabetic booklets, glucometers and insulin in the kitchen so to manage their diabetes. A (future?) requirement of the system is to be pervasive, so to make the information available where it is needed and used.

# 3 System Design

The ethnographic study (see Par. 2) identifies three strategies adopted by people in the management of the information about their health. The necessary conditions for the realization of these strategies are translated into design characteristics: flexibility, adaptability and customizability. Flexibility is the ability of a system to support individually tailored and ad hoc solutions while adaptability refers to the system capability of evolving in order to satisfy ever changing requirements. Customizability allows users to choose which functionalities can better accommodate their habits and preferences.

The sociological analysis focuses on how people organize their paper documents which are only one of the media through which health information is vehiculated. We believe that previous considerations can be safely generalized to every kind of health information. In other words, the system must be flexible and adaptable so that users can choose which kinds of information and functionalities are relevant with respect to a particular need in a period of their life.

The design of a PHR should support people in managing their health and organizing their activities. Thus we believe that a PHR architecture should be aligned to general strategies people spontaneously adopt and invent in real life for the management and organization of their own health. In other words, designing a good architecture for a PHR is more related to the aspects of social organization of work than to IT world and its buzzing words.

Nobody doubts that an IT system for a PHR must be secure, reliable and interoperable. These qualities are necessary, but not sufficient to define a PHR. In addition, smarter people have already discussed for a long time in the literature on how to make IT systems more secure, reliable and interoperable. Thus we will not deal with these technical topics here.

## 3.1 Design Principle

A design principle is an architectural paradigm for information organization that should support some design characteristics in a system. In this section we describe the design principle we followed in the definition of the system reference architecture. The underlying assumption is that the principle should drive the system towards the three design characteristics emerged from the ethnographic study.

The system is organized in three levels of abstraction. The first level describes base functionalities, that is, the bricks which can be assembled to build more complex functionalities. Base functionalities are agnostic in the sense that they do not know their context of use. For example, a web service with a CRUD (Create Read Update Delete) interface managing the list of currently taken medications implements a base functionality.

The second level is the configuration level. A configuration is a collection of functionalities which act together in order to realize a use case and respond to specific health needs. Configurations represent contexts of use. For example, the set of services for the management of a chronic disease is a configuration.

The third level is the control level. The control level defines the rules a configuration must satisfies (e.g. people privacy is a priority) and describes how evolution from a configuration to the next one is carried on. Customizability lays in this level, too.

The design principle is based on the observation of what changes more often in a PHR during the life of a person. While functionalities are quite stable, configurations are mutable because they reflect health needs which are volatile and dynamic. The modularity of a configuration makes the system more adaptable and flexible.

Coming back to examples presented in Section 2, a use case may correspond to a health problem, for example a chronic disease or aging. Health problems entail some needs such as the necessity to make information more accessible. A configuration is a collection of "tools" whose functionalities help people to sort out documents. The three strategies identified above are just requirements a configuration must satisfy to support a use case. The evolution of a disease may trigger new needs, hence components of a configuration should be substitutable with others that better fit the new context.

### 3.2 Architectural Components

The design principle is implemented at two independent levels of system architecture: presentation layer and service layer. The former has to do with human-computer interactions. The latter defines how atomic and composed functionalities are organized.

A service is a functionality indentified in and implemented by an autonomous piece of software. Services can be composed in more complex and specialized services. Services are indentified with a middle-out strategy [7] from case studies and high level functional requirements (e.g. HL7 PHR [8]), consolidated data models (e.g. PHD [9]) and from legacy systems (e.g. national health infrastructure and regional clinical document repositories).

An application is a service that presents a view of the system to humans. An application uses some services of the PHR and (possibly) some services developed by third parties. For example, if an application shows epidemic data over a country, it merges PHR services and map services of an external geo server managed by a third party. Applications can be designed also to accommodate specific needs (e.g. people affected by diabetes).

In this phase of the project, we are focusing on the development of a web application to allow an heterogeneous group of people to access basic PHR functionalities. The application follows a classical three-tier architecture and is a portal that aggregates widgets and offers some common basic functionalities (e.g. single sign-on). A widget is a self-standing application that lives within a web portal. Widgets are agnostic, that is, they do not know why and in which context they are used.

### 3.3 Underlying Data Model

The underlying data model of our PHR consists of four domains. The first domain is inspired to Project Health Design data model [9] where users are active actors in the management of their health and can generate information flows towards health providers. The domain is divided in three sub-domains: observations of daily living (e.g. glucose measurements, pain description, drug administration), medication management (e.g. list of currently taken medications) and calendaring (e.g. task and event scheduling). The second domain contains concepts modeling clinical resources and communication from and to healthcare operators; clinical records and messaging are part of this domain. The third domain captures all the concepts related to current health status and personal health history; involved concepts are close to attributes of the Continuity of Care Record. The last domain models user preferences and advanced directives.

# 4 Future Work

The ethnographic study described before gives us useful suggestions for outlining the structure of TreC prototype. The ongoing work is to design the TreC user interface so that not to compromise the user's acceptability, strongly influencing the use of the system. This section focuses on some remarks about usability of TreC user interface and older people as target user. Accessibility aspects with respect to user disabilities are out of the purpose of this section.

## 4.1 The Usability Evaluation Approach

From a conceptual point of view, usability evaluates the distance between the design model and the user model of a product. The design model describes in detail the structure of the system and how the system will be implemented. The user model refers to the operation model worked out by the user. It leads the way him/her interacts with the system. The more the design and user models of a system are close, the less the system is awkward to use. So the system interface is critical and need attention. Generally speaking, a good interface should answer for resources, limits and instructions for use; a lot of attention should be given to the relation between what a user could do using the interface layer (i.e. moving object, writing, etc.) and the following results.

The aim of a system's usability evaluation is to understand how well a system communicates with and support the functional needs of its target users. There are a lot of usability evaluation methods described in literature. Each methods can be chosen according to the usability elements we need to look into (i.e. learnability, efficiency of use, memorability, error prevention, and satisfaction) and gives better results with real user participants. Experts recommend using a combination of techniques according to research needs during the system design and development lifecycle.

## 4.2 The Target Users

Population aging is a long-range trend that characterizes affluent society. Life expectancy is longer than that in the past and there is a substantial increase in the number of older people (age 65 and older). Older people usually have more long-term health problems, chronic illnesses (such as arthritis, diabetes, high blood pressure and heart disease) than younger people. Moreover they consume more prescription and over-the-counter (OTC) drugs than any other age group. So even though TreC is designed to be a universal system, older people are the actual potential user group.

Besides this consideration, literature about universal usability issues suggests to focus specifically on the needs and functionalities of older (and disabled) people. In particular Newell et al. underline that "*in terms of their abilities, design which is appropriate for older people will be appropriate for most of the population, whereas design for younger and middle aged people will exclude significant numbers of older people*" [10]. For all this reasons we decided that older and disabled people would be the target user for the designing and development phase of TreC.

### 4.3 Implication for the System Interface

Technology could be used to approach and deal with most of the problems related to aging. In particular, smart system like TreC could empower older people in managing their health. But during the design phase we have to keep in mind the special needs of elderly people in terms of implication of age-related changes in functional abilities. In fact they face technology usability impediments related to physical, mental, and cognitive impairments. Moreover, considering the historical spread of technology and internet at work and then at home, current older people are unlikely to have experience of computer or internet use.

Usability literature make specific individual recommendations in terms of product attributes in order to face all these problems: font size, use of background and text color, etc. Nevertheless building a user interface on such recommendations is not enough. The composition of such interconnected elements may hide new usability problems that could be noticed only during a usability testing session. So in the design phase of TreC interface we have to consider the suggestions coming from the usability literature. The next step is to rebuild the interface with a pleasure graphic. Then usability testing with older people will allow to check the whole user interface. The results will be translated to appropriate input for the user interface design.

## 5 Conclusions

In this paper we have described part of a process aiming to design a PHR as patient-centered information and communication hub. We argue that the key to its success is its capability to support the existing activities carried out by laypeople in managing their health records. In this perspective, the identification of user requirements can be only elicited through the analysis of the current management strategies used by laypeople. Document management strategies are interpreted as software qualities supported by a highly modular architecture following the service-oriented paradigm.

In the next phase of our work we will devote a particular effort to early identifying usability problems that may impair the use of the TreC system. After building the user interface by giving it a pleasure graphic, the involvement of older people for carrying out interactive usability testing on refined interface versions represents an effective way for achieving this.

## Acknowledgement

This work was sponsored by the Department of Health and the Department of research and Innovation of the Autonomous Province of Trento.

## References


1. Vikkelsø, S.: Subtle Redistribution of Work, Attention and Risks: Electronic Patient Records and Organisational Consequences. Scandinavian Journal of Information Systems 17(1), 3–30 (2005)
2. Østerlund, C.: Documents in Place: Demarcating Places for Collaboration in Healthcare Settings. Computer Supported Cooperative Work 17(2-3), 195–225 (2008)
3. Connecting for Health. The personal health working group final report. Markle Foundation (2003), http://www.connectingforhealth.org/resources/final_phwg_report1.pdf (May 1, 2009)
4. Moen, A., Gregory, J., Brennan, P.: Cross-cultural factors necessary to enable design of flexible consumer health informatics system (CHIS). Int. J. Med. Inf. 76S, S168–S173 (2007)
5. Star, S.L., Straus, A.: Layers of Silence, Arenas of Voice: The Ecology of Visible and Invisible Work. CSCW 1999 8(1-2), 9–30 (1999)
6. Charmaz, K.: The grounded theory method: an explication and interpretation. In: Emerson, R.M. (ed.) Contemporary field research: a book of readings, pp. 109–126. Little, Brown, Boston (1983)
7. Josuttis, N.M.: SOA in practice. O'Reilly, Sebastopol (2007)
8. HL7 PHR System Functional Model (December 2008)
9. Sujansky & Associates LLC. PHD Common Platform Components: Functional requirements (December 2007)
10. Newell Alan, F.: Older People as a focus for Inclusive Design. 4(4), 190–199 (March 2006), http://www.gerontechjournal.net (May 1, 2009)

# Mixed-Initiative Argumentation: Group Decision Support in Medicine

Chee Fon Chang[1], Andrew Miller[2], and Aditya Ghose[1]

[1] Decision Systems Lab and Centre for Oncology Informatics
School of Computer Science & Software Engg
University of Wollongong, NSW, Australia
[2] Centre for Oncology Informatics
Illawarra Health and Medical Research Institute
University of Wollongong, NSW, Australia
{c03,amiller,aditya}@uow.edu.au

**Abstract.** This paper identifies ways in which traditional approaches to argumentation can be modified to meet the needs of practical group decision support. Three specific modifications are proposed. Firstly, a framework for accrual-based argumentation is presented. Second, a framework for outcome-driven decision rationale management is proposed that permits a novel conception of mixed-initiative argumentation. The framework is evaluated in the context of group decision support in medicine.

**Keywords:** Abstract Argumentation System, Group Decision Support, Knowledge Management System.

## 1 Introduction

In this paper, we propose a framework for mixed-initiative argumentation, which interleaves "winner determination" in the style of classical argumentation with decision identification by the user, coupled with the recording of decision rationale. We therefore present a spectrum, with classical argumentation performing "forward reasoning" model at one extreme, and decision rationale recording in the "reverse justification" model at the other extreme. Mixed-initiative argumentation represents the middle ground. The best means of obtaining a machinery for recording and managing decision rationale is to "invert" the machinery for decision generation. In other words, in a group decision making setting, we need to ask the following question: what inputs to a group decision "generation" system would have generated the selected decision? These inputs then constitute the rationale for the selected decision. Argumentation provides a basis for determining the set of "winning" arguments in settings where multiple points of view need to be accommodated, such as group decision support. Very little exists in the literature on rationale management in group decision support. Rationale management is an important question in a variety of settings. Recording decision rationale can help ensure consistency across a sequence of decision that can be

P. Kostkova (Ed.): eHealth 2009, LNICST 27, pp. 43–50, 2010.

used to justify other decisions. This is also a valuable pedagogical tool. These aspects of rationale management are critical in the medical domain.

Argumentation theory is concerned primarily with reaching conclusions through logical reasoning, starting with certain premises. Argumentation theory is therefore concerned with acceptability, and not necessarily any notion of truth or agreement. In argumentation theory, the notion of conflict is generally represented by either *attack* or *defeat* relations [1]. The use of argumentation in medical decision support is not new. In [2], the authors investigated on the collaborative decision-making and communicative discourse of groups of learners engaged in a simulated medical emergency. In [3], the authors introduced the use of arguments for decision support and advocated the need for decision support systems to support more than single, isolated decision making as most decisions are made in context of extended plans of action. However, the proposed system fails to exploit the full potential of argumentation and does not allow for rationale management or the evolution of rules and preferences. In [4], the authors provides a brief insight to a body of work centred on applications of argumentation in biomedicine. Our work differ by taking a mixed-initiative approach to elicit the background knowledge required in the group decision making as well as rationale management. To motivate our work, let us examine several extracts from a medical group decision session. The discussion involves several medical specialists (Surgeon ($S_1$,$S_2$,$S_3$), Radiation Oncologist ($RT_1$,$RT_2$)) debating on the best treatment for a patient with early stage superficial unilateral larynx cancer.

| Disease Definition: Larynx Cancer<br>Early Superficial Unilateral | |
|---|---|
| $S_1$ | : ($A_1$) My opinion is to take out the patient's larynx. This is has the best cure rate of 99%. |
| $S_2$ | : ($A_2$) I agree, taking out the patient's larynx would provide the best cure potential. |
| $S_3$ | : ($A_3$) I also agree, taking out the patient's larynx would provide the best cure potential. |
| $RT_1$ | : ($A_4$) But if you take out the patient's larynx, the patient will have no voice. |
| $RT_1$ | : ($A_5$) However, if you use radiotherapy, there is a 97% cure rate from the radiotherapy and about 97% voice quality, which is very good. The 3% who fail radiotherapy can have their larynx removed and most of these will be cured too. |

The above example illustrates several important issues. Firstly, the need for accrual in argumentation. Within argumentation, "accrual" generally refers to the grouping of arguments to support or refute a particular opinion [5]. To highlight our point, let us focus on three key arguments. $A_4$ forms the basis of an attack on the argument $A_1$. When just considering these two arguments alone, it maybe difficult to determine which course of action is the most appropriate. Now, let us consider the argument: $A_1$ in conjunction with the argument $A_5$. Again, it maybe difficult to determine which choice is a more appropriate action to take. However, when we consider all three argument together, it is clear that the best course of action is to perform radiotherapy before taking out the patient's larynx. Secondly, the ability to strengthen arguments by repetition. To illustrate our point, let us focus on the arguments: $A_1$, $A_2$, $A_3$. Although these three arguments do not enlighten the discussion with any additional information, it is conceivable that in a human debate situation, the number of arguments is sufficient enough to overwhelm any suggestion of the contrary. However, we are not advocating that we should always strengthen a position simply by providing multitude of identical arguments. Performing such task should be informed by additional information such as source's expertise or credibility. Finally, the importance of the information sources during argumentation. If we consider the accrual of identical arguments as a reflection of the norms of a community, then it is conceivable that the first course of action would be to take out the patient's larynx. However, if the specialist $RT_1$ has special insight or knowledge not shared with the others specialist (e.g. the specialist is the ONLY radiation oncologist in the group), therefore might occupy a somewhat privileged position. It is then possible that the arguments made by this particular specialist may carry more weight. In this example, we motivate that the credibility of the individual presenting the argument is important. Using this notion of credibility, we can infer a preference ordering on the arguments.

Let us consider another snippet extending from the same discussion.

---

$S_2$ : My opinion is also that the patient should have a hemi-laryngectomy. This will give a cure rate is as good as radiation therapy.

$S_3$ : I agree, performing a hemi-laryngectomy would give a cure rate as good as radiotherapy.

$RT_1$ : Yes, I have performed many hemi-laryngectomies, and when I reviewed my case load, the cure rate was 97%, which is as good as that reported internationally for radiotherapy.

$RT_2$ : I agree, however, you fail to take into account the patient's age. Given the patient is over 75, operating on the patient is not advisable as the patient may not recover from an operation.

$RT_1$ : Yes, however, in this case, the patient's performance status is extremely good, the patient will most likely recover from an operation. (i.e. the general rule does not apply)

---

Notice that the above example illustrates an interesting phenomenon. In this particular instance, the specialist $RT_1$ did not disagree with the correctness of the presented facts and conclusion in the argument presented by $RT_2$, but rather the applicability of the underlying inference rule that is used to construct the argument. This phenomenon is defined by [1,6] as "undercut". In this situation, the argument presented by $RT_1$ is more specific. This indicates that there exist some exceptions to the general decision rules that are context dependent. Furthermore, this also indicates that a revision of the general attack relation should be performed. Finally, let us consider another snippet extending from the same discussion.

---

$S_2$ : Reviewing our past case decisions, evidence suggest that the we have always performed a hemi-laryngectomy, hence my preference is to do the same.

$S_3$ : I agree, however, there is some new medical literature reporting that the voice quality after a hemi-laryngectomy was only 50% acceptable and the reporting institution was the North American leaders in hemi-laryngectomy, hence we should perform radiotherapy.

---

This example illustrates an attack on the user preference. Similar to the previous example, attacks on the user preference are generally context sensitive. This example also illustrates that an argumentation system should evolve over time, it can accumulate past decision as justification for future decisions (similar to that of a legal common law system). However, it is clear that in some instances, we wish to overrule past precedent. In most argumentation and decision support systems presented in the literature, the systems are relatively static. Most systems are open to new facts, however, have difficulties handling changing rules and preferences.

In the next section, we will present a preference-based accrual abstract argumentation framework (PAAF) modeled on works of Dung[7] and Bench-Capon[8]. Although Bench-Capon[8] presented a value-based argumentation framework, in our work, we view these values as merged preferences of the audiences. Section 3 will illustrate the use of this framework in a medical group decision system.

## 2 Formal Framework

A preference-based accrual abstract argumentation framework is a triple: $\langle AR, attacks, Bel\rangle$ where $AR$ is a set of arguments, *attacks* is a binary relation on $AR$. $Bel = \langle V, \leq, \Phi\rangle$ where $V$ is a set of abstract values, $\leq$ is a total order on $V$, $\Phi$ is a total mapping function which maps elements of $2^{AR}$ to elements of $V$. Given $\alpha, \beta \in AR$. For readability, we will denote $attacks(\alpha, \beta)$ to mean $\alpha$ attacks $\beta$. Similarly, given $v_1, v_2 \in V$, we will denote $pref(v_1, v_2)$ to mean $v_1 < v_2$ or $v_1$ is preferred to $v_2$.

Given the abstract framework, we will now define a notion of a *conflict-free* set of arguments. A *conflict-free* set of arguments is simply a set of arguments where arguments in the set do not attack each other. Let $PAAF = \langle AR, attacks, Bel\rangle$, a set of arguments $S$ is said to be **conflict-free** if and only if $\neg(\exists\alpha\exists\beta((\alpha \in S)\wedge(\beta \in S)\wedge attacks(\alpha, \beta)))$. The notion of acceptability is defined with respect to a set of arguments. An argument is acceptable to a set of arguments (in other words accepted into the set) if the set of arguments attack any arguments that attack the joining argument.

Given $PAAF = \langle AR, attacks, Bel\rangle$ and $\alpha \in AR$. $\alpha$ is **acceptable** with respect to a set of arguments $S$ (denoted as $acceptable(\alpha, S)$) if and only if $\forall\beta((\beta \in AR) \wedge attacks(\beta, \alpha)) \rightarrow \exists\gamma((\gamma \in S) \wedge attacks(\gamma, \beta)))$. Given the notion of conflict-free and acceptability, we are now able to define a notion of admissibility. Admissibility is simply defined as a set of arguments that are conflict-free and that defends itself from all attacks from arguments outside the set by attacking them. A conflict-free set of arguments $S$ is **admissible** if and only if $\forall\alpha((\alpha \in S) \rightarrow acceptable(\alpha, S))$. The admissible sets are ordered based on the $<$ ordering imposed on the abstract value $V$ and by utilising the mapping function $\Phi$, the admissible sets are assigned an abstract value.

For any given admissible set of arguments $S$, $S$ is an **preferred extension** if there is no admissible set $S' \subseteq AR$ s.t. $pref(\Phi(S'), \Phi(S))$ Note that our definition of a **preferred extension** deviates from that traditionally defined in [7,8]

Given two systems, $PAAF_1 = \langle AR, attacks_1, Bel\rangle$ and $PAAF_2 = \langle AR, attacks_2, Bel\rangle$, $attacks_1$ and $attacks_2$ are *A-consistent* if and only if $(attacks_1 \subseteq attacks_2) \vee (attacks_2 \subseteq attacks_1)$.

Given two systems, $PAAF_1 = \langle AR, attacks, Bel_1 \rangle$ and $PAAF_2 = \langle AR, attacks, Bel_2 \rangle$, where $Bel_1 = \langle V, \leq, \Phi_1 \rangle$ and $Bel_2 = \langle V, \leq, \Phi_2 \rangle$, $\Phi_1$ and $\Phi_2$ are *$\Phi$-consistent* if and only if $\nexists \alpha, \beta \subseteq AR s.t (\Phi_1(\alpha) \leq \Phi_1(\beta) \wedge \Phi_2(\beta) \leq \Phi_2(\alpha))$. The above formula allows us to detect inconsistencies between two different attack and preference relations as well as performing context sensitive revision on the attack and preference relations.

Typical usages of mixed-initiative interaction are in scenarios consisting of multiple machines cooperating or collaborating to perform tasks or activate coordination, such as, in distributed planning in multi-agent systems. Several different levels of mixed-initiative are presented in [9]. Our mixed-initiative argumentation framework falls into the "Sub-dialogue initiation" category where in certain situations the system initiate a sub-dialogue to ask for clarification which may take several interactions. Hence, the system has temporarily taken the initiative until the issue is clarified.

Our system initiates the mixed-initiative interaction by first generating a preferred extension and asking the user for verification. If the user agrees with the decision, the system terminates. If the user disagree with the generated preferred extension, a process of query and answer occurs. Requesting the user to validate arguments and provide additional arguments such that when the system recomputes, an agreement occurs. Each decision is then recorded for future reference. This is an iterative process which interleaves "winner determination" in the style of classical argumentation and decision identification by the user, coupled with the recording of decision rationale.

During the "reverse justification" interaction, any combination of three possible types of modification can occur: a new fact is introduced via a new argument, a revision on the attack relation via a new argument using the notion of a-consistency or a revision on the preference mapping via a new argument using the notion of $\Phi$-consistency. In this mode, prioritized revision occurs on the system where arguments, attack and preference relations from the user are imposed onto the system. When the user presents a set of winning arguments, four outcomes are possible during this comparison.

Firstly, the set of winning arguments identified by the user is a preferred extension with respects to the system. In this instance, no modifications are required and hence no additional arguments or rationales are required.

Secondly, the set of winning arguments is not a preferred extension with respect to the system, however, an addition of a new fact or collection of facts will allow the system to generate the identified preferred extension. In this situation, a new argument or a set of arguments is inserted. This new argument represents a reason for supporting the conclusion, hence the preferred extension constitutes a decision rationale.

Thirdly, in the case where the inconsistency lies between the attack relation, a new argument or sets of arguments that eliminate the inconsistency between the attack relations is required. To perform this, we will refer to two functions. Firstly, a function that extracts from the set of arguments the subset that is *relevant* in relation to the attack relation. In particular, the subset of arguments that makes reference to the particular attack relation within its sub-structure. Thus, $rel_{attacks}(S)$ represents that subset of a set of arguments $S$ that make reference to the offending *attacks* relation. Secondly, a function that will extract from the sub-structure of each argument the encoded attack relation. Thus, $ext_{AR}(S)$ represents a set of binary relations extract from the set of arguments $S$ with respect to some set of arguments $AR$. For example, assume that we have a set of arguments $AR = \{\alpha, \beta\, \gamma\}$, an attacks relation consisting of $\{attacks(\alpha, \beta)\}$ and a set of arguments $S = \{\alpha, \beta, \delta\}$ and within the substructure of $\delta$ the attack relation $\neg attacks(\alpha, \beta)$. $ext_{AR}(rel_{attacks}(S))$ will return $\{\neg attacks(\alpha, \beta)\}$. We will then augment the existing attack relation with that extracted from the arguments.

Similarly, in the case where the inconsistency lies between the preference mapping function, a new argument or sets of arguments that eliminate the inconsistent between the mapping function is required. For any specific instance, we eliminate the inconsistencies between mapping functions by substituting the existing mapping function with one that is encoded in the arguments. We will again refer to two functions. Firstly, a function that extracts from the set of arguments the subset that is *relevant* in relation to the preference mapping function. In particular, the subset of arguments that makes reference to the mapping function within its sub-structure. Thus, $rel_{\Phi}(S)$ represents that subset of a set of arguments $S$ that make reference to the mapping function. Secondly, a function that will extract from the sub-structure of each argument the encoded mapping function. Thus, $ext_{\Phi}(S)$ represents a set of binary relations extract from the set of arguments $S$ with respect to some set of arguments $\Phi$. We will then replace the existing preference mapping function with that extracted from the arguments. Two key benefits exist in such an approach. Firstly, we are able to evolve an existing argumentation rather than constructing a new argumentation system with new attack and preference relations. This allows for the reuse of arguments, attack and preference relation. Secondly, we are able to address "traceability" issues as the system accumulates justifications as it evolves from one instance to the next, hence allowing us to manage rationale over the life of the system. This allows the process of argumentation to form the basis for rationale management.

## 3 Medical Group Decision Support System

Utilising a Web 2.0 philosophy, we have constructed a web enabled medical group decision support system utilising Asynchronous JavaScript and XML (AJAX) with a back-end repository. HyperText Markup Language (HTML) and Javascript are used to build the user interface and controls the interaction with the web server. Hypertext Preprocessor (PHP) is used to build the reasoning engine to perform back-end computation of the arguments. MySQL is used as the database repository. The benefits of this appraoch are platform independence, portability, scalability and accessibility.

The prototype was presented to several oncologists and a "head-and-neck" session was simulated. A "head-and-neck" session is where groups of oncologists meet to discuss treatment therapy for cancer cases in the head to neck region. During this session, a typical larynx cancer case was discussed. Treatment analyses are performed over 5 categories. These categories (in order of importance) are as listed: survival, control, physical toxicity, psychological toxicity and clinician's choice. These categories are addressed in stages starting from the most important to the least. Argumentation is performed at each stage and final recommendation is based on the accrual of all arguments over all the stages. Each stage can be viewed as a decision-making cycle where decision made affects the available choices for the next cycle. Given a case description, the system presents a possible recommendation (if one exists). Specialists are then asked if the recommendation is acceptable. If the recommendation is not acceptable, the system asks the specialist to select a recommendation and justify it with arguments, with which the system then recomputes a new recommendation. If the recommendation does not coincide, the system presents its findings and asks for more justifications. This process is iterated until the recommendation of the system coincides with the specialist's choice.

In Figure 1, we present the argument modification interface. The users are allowed to add, delete and modify the arguments associated with a particular treatment choice in the forward learning mode. The changes require that the clinician provide the strength of the evidence and a literature references if used. In essence, by associating the argument with the treatment choice, the user has provided justification for the particular treatment.

In Figure 2, we present the resulting output, which illustrates the recommended decision for each facet of a given sequence of decisions. Each facet has different priority (if two treatments have identical cure and control rates, the one with lower physical toxicity is preferred) and the final treatment choice is computed using these preferences. This figure also illustrates the ability for the user to validate the recommendations and subsequently activate the second of the two learning modes where the user disagrees with the recommendation.

**Fig. 1.** Arguments

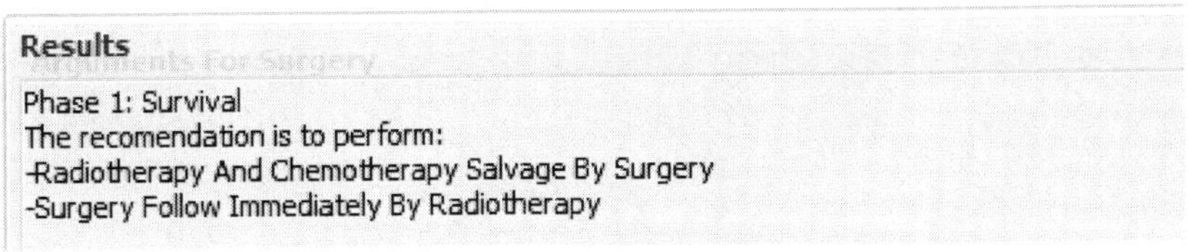

**Fig. 2.** Recommendations

The general response to using the tool has been positive. The specialists found that the tool is useful both as a practical tool for a trained specialist as well as a teaching aid for medical registrar. The specialists also felt that the tool has huge potential, however, since the tool is still in its' infancy there are still several practical issues that needs addressing. During the trial, several issues were identified. These issues falls into two categories: usability of the user interface and performance of the argumentation engine.

During an original execution of the tool, we found that when computing recommendations, the tool can take sometime to return a decision (in some cases, several hours). This issue was address by limiting the scope to the available 9 therapy choices rather than allowing the system to compute all therapy choices (including non-existing ones).

## 4 Conclusion

We have identified ways in which traditional approaches to argumentation can be modified to meet the needs of practical group decision support, presented a tool and described its evaluation in the context of group decision support for medical oncology.

## References

1. Prakken, H., Vreeswijk, G.: Logics for Defeasible Argumentation, 2nd edn. Handbook of Philosophical Logic, vol. 4, pp. 218–319. Kluwer Academic Publishers, Dordrecht (2002)
2. Lu, J., Lajoie, S.P.: Supporting medical decision making with argumentation tools. Contemporary Educational Psychology 33(3), 425–442 (2008); Collaborative Discourse, Argumentation, and Learning
3. Glasspool, D., Fox, J., Castillo, F.D., Monaghan, V.E.L.: Interactive decision support for medical planning. In: Dojat, M., Keravnou, E.T., Barahona, P. (eds.) AIME 2003. LNCS (LNAI), vol. 2780, pp. 335–339. Springer, Heidelberg (2003)
4. Fox, J., Glasspool, D., Grecu, D., Modgil, S., South, M., Patkar, V.: Argumentation-based inference and decision making–a medical perspective. IEEE Intelligent Systems 22(6), 34–41 (2007)
5. Verheij, B.H.: Reason based logic and legal knowledge representation. In: Carr, I., Narayanan, A. (eds.) Proceedings of the 4th National Conference on Law, Computers and Artificial Intelligence, University of Exeter, pp. 154–165 (1994)
6. Pollock, J.L.: Defeasible reasoning. Cognitive Science 11(4), 481–518 (1987)
7. Dung, P.M.: An argumentation-theoretic foundation for logic programming. Journal of Logic Programming 22(2), 151–171 (1995)
8. Bench-Capon, T.J.M.: Persuasion in practical argument using value-based argumentation frameworks. Journal of Logic and Computation 13(3), 429–448 (2003)
9. Allen, J.F.: Mixed-initiative interaction. IEEE Intelligent Systems 14(5), 14–23 (1999)

# Enabling Technology to Advance Health-Protecting Individual Rights-Are We Walking the Talk?

Crystal Sharp and Femida Gwadry-Sridhar

I-THINK Research, Lawson Health Research Institute, 268 Grosvenor St, FB-112, London, ON N6A 4V2 Canada
crystal.sharp@lawsonresearch.com

**Abstract.** The evolving structure and business of health care services and delivery need the functionality and capability offered by electronic health record (EHR) systems. By electronically diffusing the traditional patient record, however, this new model blurs the long-established medical data home, raising concerns about data ownership, confidentiality, access and individual rights. In 2008 the Lawson Health Research Institute began the process of instituting a robust health informatics and collaborative research infrastructure, now known as I-THINK Research. As data are migrated to the platform and policies are developed, we are forced to confront the complexity of issues around protection of individual rights. The paper presents, in a broader context, the main issues surrounding the privacy debate and the need for education, accountability and new legislation to help define and protect individual rights as new e-health business models emerge.

**Keywords:** eHealth, electronic health records, consent, Google Health, Microsoft Vault, personal health records, privacy, confidentiality.

## 1 Introduction

In 2008 the Lawson Health Research Institute (the research institute for the London Health Sciences Centre and St. Joseph's Health Care in London, Ontario, Canada) began the process of instituting a robust health informatics and collaborative research infrastructure, now known as I-THINK Research[1]. Our ethos is that a secure and common platform for rich clinical, biomarker, genomic and imaging data, that otherwise exists in silos, will enable collaborative research yielding deeper insight into disease etiology and better care strategies for patients. As data are migrated to the platform and policies are developed, we are forced to confront the complexity of issues around protection of individual rights. This paper discusses some of the complexities but in a broader context because as health researchers, practitioners and consumers we are all affected by rapidly evolving use of health technology, new business models, privacy legislation and definitions of ownership, sharing and management of personal data.

It is a surprising fact, given the rapid adoption of computerization in business and personal use, that doctors in most parts of the world still work mainly with pen and paper[2]. Even in an advanced industrialized economy like the USA, a recent study

P. Kostkova (Ed.): eHealth 2009, LNICST 27, pp. 51–61, 2010.

found that only 1.5% of acute care hospitals in the American Medical Association had a comprehensive electronic-records system (present in all clinical units), an additional 7.6% had a basic system (present in at least one clinical unit); and computerized provider-order entry for medications were implemented in only 17% of hospitals[3]. The 2009 American Recovery and Reinvestment Act has recently allocated up to $20 billion to implement clinical information systems, anticipating use of electronic health information for each person in the United States by 2014[2]. Given the current low levels of adoption of electronic health records in U.S, however, the combination of financial support, interoperability, and training of technical support staff needed to spur adoption of e-health will be considerable[3],[4]. Public policy challenges of implementation will require improvements to existing law, new rules for entities outside the traditional health care sector, a more nuanced approach to the role of consent, and stronger enforcement mechanisms[5]. With rare exception, however, national efforts to advance health information technology have not adequately addressed privacy[5]. The debate about rights to privacy has often seemed too polarized to resolve, grid locking initiatives to promote health information technology. It will be interesting to see how this initiative unfolds.

This paper will present the main issues surrounding the privacy debate in context of the potential of e-health to improve health care, the difficulties in defining ownership of data in a networked world, and the need for education, accountability and new legislation to help define and protect individual rights as new e-health business models emerge.

## 2 Potential of E-Health

The evolving structure and business of health care services and delivery need the functionality and capability offered by electronic health record (EHR)systems, which offer the potential to[6]: facilitate accurate and efficient flow of patient medical and billing information between organizationally and geographically distinct providers; to enable utilization review; to encourage patient-centred care, by giving patients access to information in their records, particularly when they have fragmented or episodic relationships with multiple providers; and to give providers transparent access to other occasions of treatment, particularly pharmacotherapy. Both patients and regulators want increasing amounts of data regarding errors or near misses and outcomes in populations—data that is difficult to generate without sophisticated data coding and nearly impossible to analyze without complex, comprehensive database systems.

By electronically diffusing the traditional patient record, however, this new model blurs the long-established medical data home, raising concerns about data ownership, confidentiality, access and individual rights. Since electronic health records offer the potential for high-quality, interactive and innovative patient-centred care, it is imperative that such issues are resolved if their full potential is to be realized[7].

## 3 Difficulty in Defining Ownership of Data

Networked ICT applications in energy, transportation, health and digital libraries, critical to the well-being of humankind, are currently being developed and delivered via the internet, widening social, economic, and productivity boundaries and creating new value chains[8]. E-health infrastructure upkeep challenges involve high-capacity storage, secure systems, and connectivity for different devices requiring high bandwidth, both wired and wireless. In addition to the hospitals, the care providers, consumers, researchers, insurers and businesses that cater to the needs of healthcare delivery, there are pharmaceutical companies, health systems vendors, medical device producers and data brokers. EHR systems are becoming more and more sophisticated and nowadays include numerous applications, which are not only accessed by medical professionals, but also by accounting and administrative personnel. This could represent a problem concerning basic rights such as privacy and confidentiality. By some views, granting access to an EHR should be dependent upon the owner of the record to define who is allowed to access. But who owns medical information – the one who gives care, receives care, pays for care?[9]. Ownership of paper records was never much in doubt. Clinicians and insurers own the tangible vessels in which they store patients' medical information. But now that digitizing information frees it from particular storage media, confusion reigns. By some views the rights of ownership should belong to the patient[9;10].

Determining ownership of health information is not straightforward and modern health information systems elevate issues of ownership. Health information data may carry personal property rights, with numerous stakeholders making cognizable claims to ownership. The rights to a health information system involve the ascertainment of intellectual property. Both types of property rights - personal and intellectual - are immensely valuable and thus the definition of ownership is important[11],[12]. Many of the legal questions that arise relate to the formation of the information industry. Others, including some of the most difficult-to-resolve issues, involve complex balancing questions. Who should have access to certain types of information? Should industry members enjoy protections against certain types of liability for conduct that may become more evident as the volume of information expands? In other words, should the law recognize certain privileges, “safety zones,” and “safe harbors,” in exchange for the creation, collection, sharing, and use of certain valued information, and if so, what conditions should be placed on sanctioned activities? Finally, are there extreme cases in which the government should compel, rather than merely incentivize and encourage, the creation and production of certain information?[11].

## 3.1 The Complexities of "Privacy" in a Networked World

Patient concerns about the privacy, confidentiality and security of their health data are legitimate, as disclosure about existing health conditions may affect an individual's employment, ability to get and maintain health coverage without having to pay a high premium. Policy considerations to safeguard privacy amount to balancing individual privacy considerations against the accessibility of health information; delineating the required and permissible purposes for which vast new amounts of health information can be collected, stored, transmitted, used and disclosed; and negotiating the regulatory boundaries of new technologies[11;13].

The effective coordination of health care relies on communication of confidential information about consumers between different health and community care services. The legal and ethical issues involve consent and its alternatives, the handling of identifiability. Matching and combining data from multiple databases, especially at the individual level, is a powerful tool made possible by the availability of high performance computational power and rich databases. Data linking raises a number of issues. If the data are of sufficient richness to enable identification of individuals simply by adding the various factors such as education, profession, marital status etc. together to reach an almost certain conclusion with regard to the identity of the person, then this is a direct violation of privacy and data protection legislation. Linked-up material does amount to a fuller description than the bits unlinked, and thus may present higher potential for abuse[14-17]. And in the case that the component data are not identified, interlinking them may provide more cues and decrease the difficulty of re-identifying the subjects by deduction. Whether some degree of linkage may be "too much" relative to the benefits and safeguards has to be judged in context[17]. Mobile sensor technologies that gather data ubiquitously and unobtrusively present a whole host of new security and privacy concerns[18].

Distinguishing between the unique needs of information-based research, which uses medical records or stored biological samples, and interventional clinical research, which involves people who participate in experimental treatment, the Institute of Medicine, HIPAA committee, recommends extending the Common Rule (a set of federal regulations for research involving human subjects that requires a review of proposed research by an Institutional Review Board (IRB), the informed consent of research subjects, and institutional assurances of compliance with the regulations) to apply to all interventional research, regardless of funding source[19]. However, Research and Ethics Boards typically do not have the necessary expertise to assess electronic health privacy[20] and this can be a problem.

### 3.2 Privacy Legislation

The law itself can both advance change and impede it – this is particularly evident in health care. Legal complexities arise as a function of policymakers' efforts to balance the rights of patients and their expectations of privacy; the autonomy and authority of health professionals; market-based economies dominated by the buying and selling of health care; and the delicate balance of jurisdictional powers over health care quality, financing and accountability. No aspect of health care offers a better example of the challenges inherent in balancing these interests than the collection, management, disclosure and reporting of health information [11].

Privacy laws (the Canadian Personal Information Protection and Electronic Documents Act [PIPEDA], US Health Insurance Portability and Accountability Act [HIPAA], and European Union Data Privacy Directive, for example) generally define personal information as *identifiable* information *about* and individual and require that individual's consent before such personal information can be collected, used or disclosed, in the absence of some applicable exception[16].

The concept of informed consent itself is fraught with complications. For instance it is debatable whether a child or a mentally incapacitated individual is adequately informed to be able to consent. Research that is based on retrospectively collected records may never be allowed to commence if subjects cannot be located for their consent. Researchers have legitimate concerns about completeness and validity in sampling and systematic bias in research results if potential research subjects can opt out[21].

The two poles in the consent argument represent a tradeoff between high coverage of the population, where records can be uploaded without explicit consent, and the opposite situation where explicit consent must be obtained for every record uploaded. As mentioned by Greenhalgh et al (2008), shared electronic record programmes in Scotland (emergency care summary), Wales (individual health record), and France (dossier médical personnel) have to some extent squared this circle by combining "implied consent to upload" with "explicit consent to view" at the point of care, although they have not been without controversy [22].

Although de-identification is a crucial protective strategy in privacy legislation, some entities are not covered under existing privacy legislation. For example, in Canada prescription data is routinely sold or transferred to commercial data brokers who may process the data and re-sell it to pharmaceutical companies in the form of prescribing patterns or practices[15;16;23]. Thus potential patient identifiers and physician-linked prescription data "stream from pharmacy computers via commercial compilers to pharmaceutical companies" without the informed consent of patients or of physicians[23].

Reversible anonymisation, or key-coding, which maintains a connection between substantive data and personal identifiers but does not allow researchers to know the identifiers, could serve both privacy and research well[17]. Properly anonymised data are not "personal," so their processing is not generally regulated by data protection legislation. But anonymisation has its difficulties – because identifiability is a continuum and anonymisation is rarely absolute, and because there can be many reasons for retaining the potential to re-identify data[17]

# 4 New Business Models – Need for Education, Accountability and New Legislation

## 4.1 Personalized Medicine

Biobanks (repositories of tissue and DNA samples) yield data that can be linked to personal medical information and test results which in combination can provide insights into disease progression that tissue samples or medical records alone cannot. Drug companies and medical researchers can, for example, pick out samples from people with a particular disease and determine its associated genetic variations to aid drug discovery. Public-health officials and epidemiologists can identify disease patterns in subpopulations and ethnic groups far more quickly than has been possible in the past. Disease-specific biobanks have potential to accelerate research into disorders such as AIDS and breast cancer[24].

In 2005 Britain and Norway announced a plan to co-operate on biobank-based research into the causes of attention-deficit hyperactivity disorder (ADHD), autism, schizophrenia and diabetes. Norway was collecting blood samples and health data from 200,000 citizens and from 100,000 pregnant women. Britain's project, UK Biobank, began gathering blood and urine samples and confidential lifestyle data from 500,000 volunteers aged 40-69, in an attempt to untangle the genetic and environmental causes of heart disease, Alzheimer's, diabetes and cancer. Participants will provide new samples and data for up to 30 years, allowing the development and course of different diseases to be tracked. Similarly, the Karolinska Institute in Stockholm which runs one of the world's oldest university-based biobanks is following 500,000 Swedes for 30 years to gain new insights into depression, cancer and heart disease. Other national biobank projects include the Estonian Genome Project, Singapore Tissue Network, Mexico's INMEGEN, and Quebec's CARTaGENE. Various university medical schools around the world have been collecting biological samples and clinical data as a matter of routine. These resources, if shared, could now turn out to be extremely valuable for disease discovery[24].

Maintenance of patient confidentiality is a major challenge in a networked environment, because the combination of clinical data and personal data and the place of research can be enough to reveal a research participant's identity[25;26]. Britain's UK Biobank, for example, encrypts the identity of donors, so that only selected users are able to link samples and data to particular individuals. Total anonymity, however, raises problems of its own: it precludes the possibility of informing donors or their relatives if donated material reveals them to be at risk from a specific disease[24].

If a risk of identification remains, patients should be asked for consent to data sharing as well as consent to taking part in the research. The question of confidentiality is bound up with another conundrum: who is going to pay for data storage and maintenance in health grids and biobanks? The answer is unclear. One approach would be to make information freely available to academic and government researchers, but to charge drug companies and other commercial interests which stand to profit from their use of the data. That could make biobanks self-sustaining, or even profitable; it has even been suggested that donors should be given a share of the proceeds. Advances in data-mining technologies and a growing interest in the notion of "personalised" medicine have spurred a growing realisation, in both the health-care and information-technology sectors, that biobanking could be very lucrative[24]. Purists insist that biobanks should remain strictly non-commercial entities. The Genome Institute of Singapore forbids any commercialisation of its biobank data, for example, though so far it is the exception to the rule[24].

## 4.2 Personal Health Records and Personal Medical Monitoring

Google Health, released in May 2008, and Microsoft HealthVault, launched in October 2007, allow consumers to store and manage their personal medical data online. Users are now able to gather information from doctors, hospitals, and testing laboratories and share it with new medical providers, making it easier to coordinate care for complicated conditions and spot potential drug interactions or other problems. Both Google and Microsoft also offer links to third-party services that provide medication reminders and programs that track users' blood-pressure and glucose readings over time. What Google and Microsoft promise to do with electronic records is also a radical departure, both conceptually and in practice. Currently, patients who have electronic access generally use portals maintained by doctors or health-care systems. Typically, patients can view information such as prescriptions, lab results, and diagnoses; sometimes they can e-mail doctors or make appointments online. In most cases, though, patients do not control their own data, so they cannot transfer it electronically to a different health-care provider or plug it in to third-party applications[27]. With HealthVault and Google Health, however, consumers have fundamental ownership of their medical data, much as they do with financial records. As more health-care providers begin participating, it will be easy for patients to share CT scans, x-rays, and lab results with new doctors[27].

Home medical monitors, such as those for blood pressure, have become a common presence in personal health-care. Wireless technologies and web portals offer new ways to track and store the information gathered by such monitors online, which can make it easier for people to review and share test results. Microsoft HealthVault offers options for tracking health measures at home. Data from these devices can be uploaded manually or automatically via wireless sensors directly into a patient's HealthVault record, where users can then create a handy graph of their blood pressure, weight, blood sugar, or other data, and share it with their doctors or family members[28]. The Mayo Clinic recently launched a free software program, available to anyone, that piggybacks on HealthVault, integrating health history and data from medical monitors and providing reminders about vaccinations and other preventative measures. The Cleveland Clinic started a pilot program using HealthVault in conjunction with different devices to manage three chronic conditions, including diabetes, heart failure, and hypertension. Scientists will track how effective the system is at changing both treatment and patient outcomes. Systems like these enable researchers to gather information in near real time and to act on the results of that information in a more continuous fashion [28]. In the long term, HealthVault is expected to function more as a database for storing data, while third-party applications can help patients organize and act on it[28].

Currently, if you are a Google account holder, you can set up access to Google Health and enter your own medical information and even search your prescription history with a few big pharmacies. In May 2008, Beth Israel Deaconess Hospital joined the Cleveland Clinic to become Google's first partners in the new service, along with and a handful of pharmacies, labs, and other health businesses. If Google Health succeeds at Beth Israel Deaconess, this may forecast whether patients are willing to trust their health information to large personal health record (PHR) providers, and it may hint at how Google Health and similar services might impact medical care in the future[29]. Dr. John Halamka, chief information officer at Beth Israel Deaconess, Chair of the national Health Information Technology Standards Panel, and member of Google Health's advisory council, is a strong believer that patients should be the stewards of their own medical data"[29]. But it is not clear that this view would be universally shared by the medical profession and other important stakeholders.

## 4.3 Need to Provide Education and Technical Assistance for Consumers

A 2009 study using observational and narrative data to examine the acceptability, adoption and use of personally controlled health records found low levels of familiarity with PCHRs along with high expectations of the capabilities of nascent systems –a potentially problematic pairing[30]. Perceived value for PCHRs was highest around abilities to co-locate, view, update and share health information with providers. Expectations were lowest for opportunities to participate in research. Early adopters perceived that PCHR benefits outweighed perceived risks, including those related to inadvertent or intentional information disclosure. Endorsement of a dynamic platform model PCHR was evidenced by preferences for embedded searching, linking, and messaging capabilities in PCHRs; by high expectations for within-system tailored communications; and by expectation of linkages between self-report and clinical data. The author advocated educational and technical assistance for lay users and providers as critical to meeting challenges of access to PCHRs (especially among older cohorts); workflow demands and resistance to change among providers; health and technology literacy; clarification of boundaries and responsibility for ensuring accuracy and integrity of health information across distributed data systems; and improving understanding of confidentiality and privacy risks[30].

### 4.4 Need for Methods to Validate and to Ensure Accountability

Today, many journals are asking authors to include a data sharing statement at the end of each original research article. The statement is required to explain which additional data – if any – are available, to whom, and how. The data can range from additional explanatory material to the complete dataset. Those allowed access to the data might be restricted to fellow researchers only or could include everyone. Data could be available only on request, accessible online with a password, or openly accessible to all on the web with a link. Sharing could allow other researchers –and perhaps scientists, clinicians, and patients, access to raw numbers, analyses, facts, ideas, and images that do not make it into published articles and registries[31]. Potential benefits include quicker scientific discovery and learning, better understanding of research methods and results, more transparency about the quality of research, and greater ability to confirm or refute research through replication. However, such sharing also raises important questions about who owns the data[25] who gives permission to release the data (including funders, research participants, owners of the intellectual property, and copyright holders), where and how the data should be stored (in electronic repositories managed locally, nationally, or internationally; or in subject specific databases), how the data should be stored and managed and made compatible across repositories, how the data should be accessed and mined, who should have access and when, and what limits may be needed to prevent misuse and mishandling of data[31].

### 4.5 Need for Legislation That Covers All Entities

Existing health data privacy legislation (like HIPAA and PIPEDA) does not cover entities like Google Health and Microsoft Healthvault as they are not healthcare provider organizations. Dr. Halamka, CIO Beth Israel Hospital, a strong supporter of PCHRs, is of the opinion that since Google Health and Microsoft Vault monetize these sites by attracting search traffic, they would be highly motivated to build secure and trustworthy systems. However, there certainly seems to be a need for a privacy protection framework that can be applied to all PCHR products - those tethered to an EHR, those offered by a payer, those sponsored by an employer or those created by third party vendor ensures that consumers have a rubric to evaluate these products[32].

## 5 Conclusion -- Enabling Technology to Advance Health - Protecting Individual Rights - Are We Walking the Talk?

A cross sectional study in New Zealand examining the public's perception of the security of electronic systems concluded that for the EHR to be fully integrating in the health sector, there are two main issues that need to be addressed: the security of the EHR system has to be of the highest level and be constantly monitored and updated; and the involvement of the health consumer in the ownership and maintenance of their health record needs to be more proactive. The results from this study indicated that the consumer is ready to accept the transition as long as one could be assured of the security of the system[33].

Within the rapidly changing internet environment, the playing field is indeed global, raising questions about the role of government. In a discussion during the ICT 2008 conference in Lyon, France, the consensus was that government's role in this environment should not be to establish the "grand design"; but to build confidence in the system so that a user asking the following questions can be reassured: "Is my money safe? Is my data secure and being used for the purpose for which it was collected? Is my privacy protected?"[8]. The future of the Internet must be planned and pursued by considering its circular interaction with social and physical-world processes. With all the complexities presented in the networked environment, perhaps control of privacy should be shifted to consumers, allowing them to control the "privacy dial," since we all have varying levels of comfort with openness. Rather than have blanket rules for privacy protection, consumers can be given the tools to set the privacy dial to their level of comfort.

The good news is that the issues presented above are being addressed and debated. Rand Europe has recently reviewed and provided recommendations for addressing limitations in the existing European Data Protection Directive[34] as has the Data Protection Working Party[35]. Led and operated by the Markle Foundation, Connecting for Health is a public-private collaborative that includes representatives from over 100 organizations, comprising a diverse group of health care stakeholders. Connecting for Health has developed a Common Framework for Networked Personal Health Information to address the key challenges. The framework proposes a set of practices that, when taken together, encourage appropriate handling of personal health information as it flows to and from personal health records and similar applications or supporting services[36].

These are steps in the right direction. However, the complexity of technological, societal, economic, stakeholder, individual and political issues in the privacy debate make enforcement of rules extremely difficult, if not impossible. The public sector has a role in encouraging education frameworks that develop innovation and management excellence and in providing a supportive environment for entrepreneurial activity that promotes knowledge transfer and research uptake[8]. Patient-controlled health records offer one potential solution to many of the problems encountered in creating institution-based longitudinal medical records. No matter which path is taken, however, clear but adaptable laws are needed so that stakeholders can assign economic value to the access, control, and use of the medical information contained in electronic health record networks[9]. Until then we will have to be watchful that, at least in our own spheres of operation, we achieve a reasonable balance in encouraging research that can improve health while respecting individual rights.

## References


[1] Mann, R., Gwadry-Sridhar, F., Bowman, S., Soer, J.: How to Develop a Common Platform to Enable Interdisciplinary Research. International Journal of Technology, Knowledge and Society 5, 21–38 (2009)

[2] Steinbrook, R.: Personally Controlled Online Health Data – The Next Big Thing in Medical Care? N. Engl. J. Med. 358, 1653–1656 (2008)

[3] Jha, A.K., DesRoches, C.M., Campbell, E.G., Donelan, K., Rao, S.R., Ferris, T.G., Shields, A., Rosenbaum, S., Blumenthal, D.: Use of electronic health records in U.S. hospitals. N. Engl. J. Med. 360, 1628–1638 (2009)

[4] Economist, Medine goes digital: A special report on healthcare and technology. Economist, 1–16 (2009)

[5] McGraw, D., Dempsey, J.X., Harris, L., Goldman, J.: Privacy As An Enabler, Not An Impediment: Building Trust Into Health Information Exchange. Health Aff. 28, 416–427 (2009)

[6] Lang, R.D.: Blurring the lines: who owns the medical data home? J. Healthc. Inf. Manag. 22, 2–4 (2008)

[7] Gunter, T.D., Terry, N.P.: The emergence of national electronic health record architectures in the United States and Australia: models, costs, and questions. J. Med. Internet. Res. 7, e3 (2005)

[8] Sharp, C.: Conference Report: ICT 2008 - I's to the Future: Invention, Innovation, Impact. Online 33[March/April 2], pp. 22–25. Information Today, Inc. (2009)

[9] Hall, M.A., Schulman, K.A.: Ownership of Medical Information. JAMA 301, 1282–1284 (2009)

[10] Falcao-Reis, F., Costa-Pereira, A., Correia, M.E.: Access and privacy rights using web security standards to increase patient empowerment. Stud. Health Technol. Inform. 137, 275–285 (2008)

[11] Rosenbaum, S., Painter, M.: Assessing Legal Implications of Using Health Data to Improve Health Care Quality and Eliminate Health Care Disparities. The Robert Wood Johnson Foundation (2005)

[12] Bluml, B.M., Crooks, G.M.: Designing solutions for securing patient privacy–meeting the demands of health care in the 21st century. J. Am. Pharm. Assoc. (Wash.) 39, 402–407 (1999)

[13] Conn, J.: Data encryption just one option under security law. Modern Healthcare (2009)

[14] Kelman, C., Bass, A., Holman, C.: Research use of linked health data - a best practice protocol. Australian and New Zealand Journal of Public Health 26, 251–255 (2002)

[15] El Emam, K., Kosseim, P.: Privacy Interests in Prescription Data, Part 2. IEEE Security & Privacy, 75–78 (2009)

[16] Kosseim, P., El Emam, K.: Privacy Interests in Prescription Data, Part 1. IEEE Security & Privacy 72 (2009)

[17] Lowrance, W.W.: Learning from experience: privacy and the secondary use of data in health research. J. Biolaw. Bus. 6, 30–60 (2003)

[18] Nixon, P., Wagealla, W., English, C., Terzis, S.: Security, privacy and trust issues in smart environments, Glasgow, Scotland, The Global and Pervasive Computing Group, Department of Computer and Information Sciences, University of Stathclyde (2009); 5-18-0090

[19] Committee on Health Research and the Privacy of Health Information. Beyond the HIPAA Privacy Rule: Enhancing Privacy, Improving Health Through Research: Report Brief, Washington DC, Institute of Medicine (2009)

[20] Lysyk, M., El Emam, K., Lucock, C., Power, M., Willison, D.: Privacy Guidelines Workshop Report, Ottawa, Canada (2006) 7-6-0090
[21] Sharp, C.: Electronic Health Information: A boon and a curse! The Free Pint, Newsletter (2001); 7-6-0090
[22] Greenhalgh, T., Stramer, K., Bratan, T., Byrne, E., Mohammad, Y., Russell, J.: Introduction of shared electronic records: multi-site case study using diffusion of innovation theory. BMJ 337, a1786 (2008)
[23] Zoutman, D.E., Ford, B.D., Bassili, A.R.: The confidentiality of patient and physician information in pharmacy prescription records. CMAJ 170, 815–816 (2004)
[24] Economist. Medicine's new central bankers. Economist (December 8, 2005)
[25] Vickers, A.: Whose data set is it anyway? Sharing raw data from randomized trials. Trials 7, 15 (2006)
[26] Hrynaszkiewicz, I., Altman, D.G.: Towards agreement on best practice for publishing raw clinical trial data. Trials 10, 17 (2009)
[27] Schaffer, A.: Your Medical Data Online: Google and Microsoft are offering rival programs that let people manage their own health information. Technology Review (July/August 2008)
[28] Singer, E.: Personal Medical Monitoring: Keeping tabs on your vitals with Microsoft HealthVault. Technology Review (April 24, 2009)
[29] Harris, L.: Google Health Heads to the Hospital: A new partnership at a Boston hospital could forecast future success. Technology Review (May 28, 2008)
[30] Weitzman, E.R., Kaci, L., Mandl, K.D.: Acceptability of a personally controlled health record in a community-based setting: implications for policy and design. J. Med. Internet. Res. 11, e14 (2009)
[31] Groves, T.: Managing UK research data for future use. BMJ 338 (2009)
[32] Halamka, J.: Blog Entry: A Privacy Framework for Personal Health Records, December 17. Blog (2008)
[33] Chhanabhai, P., Holt, A.: Consumers are ready to accept the transition to online and electronic records if they can be assured of the security measures. Med. Gen. Med. 9, 8 (2007)
[34] Robinson, N., Graux, H., Botterman, M., Valeri, L.: Review of the European Data Protection Directive. TR7 10-ICO, Cambridge, UK, Rand Europe (2009)
[35] Halliday, D., Dizon, M., Kemmitt, H.: Baker & McKenzie's regular article tracking developments in EU law relating to IP, IT and telecommunications. Computer Law and Security Report 23, 227–232 (2007)
[36] Connecting For Health. Common Framework for Networked Personal Health Information. Connecting for Health Website (2009); The Markle Foundation, 7-6-0090

# Detecting Human Motion: Introducing Step, Fall and ADL Algorithms

Dries Vermeiren, Maarten Weyn, and Geert De Ron

Artesis University College of Antwerp, Antwerp, Belgium
dries.vermeiren@artesis.be, maarten.weyn@artesis.be,
geert.deron@ieee.org
http://www.e-lab.be/

**Abstract.** Telecare is the term given to offering remote care to elderly and vulnerable people, providing them with the care and reassurance needed to allow them to keep living at home. As telecare is gaining research interests, we'll introduce a system which can be used to monitor the steps, falls and daily activities of high risk populations in this paper. Using this system it is possible for a patient to rehabilitate at home or for elderly to keep living independently in their own house while they are still monitored. This leads to a huge cost reduction in health services and moreover it will make patients satisfied for being able to live at home as long as possible and in all comfort.

**Keywords:** MEMS, Step Detection, Fall Detection, ADL, Freescale.

## 1 Introduction

Elderly people are the fastest growing segment of the population. Due to Europe's population pyramid, aging people are becoming a point of interest even faster than in the rest of the world. In 2035, one third of the Europeans will be more than over 65 years old, which will result in huge strains on health care services [5]. This will also cause a serious social and financial problem. Care centers will have to deal with a lack of rooms and cost reduction will become one of the most important objectives in public health services [5]. One of the possible solutions is to comply with the wish of elderly to keep living at home independently as long as possible. As we do this, an increasing number of high risk populations will be living alone at home. Therefore new advanced monitoring systems are gathering more research popularity. Not only for the elderly, but also for other high risk populations (people su_ering from illnesses such as epilepsy or Alzheimer or in the case of recent surgical intervention) will long-term monitoring become an issue.

In this paper we will describe a system based on 2 tri-axial accelerometers to detect the Activities of Daily Living (ADL) of a patient and to detect its steps and falls. In the first part we will focus on the most appropriate position of the sensor on the patient's body in order to receive clear signals of their movements. Furthermore we will describe different methods for detecting steps with techniques based on simple filters and thresholds or templates. Next to this we will also study the different methods to detect falls and a very basic ADL detection method is proposed. Subsequently we will

P. Kostkova (Ed.): eHealth 2009, LNICST 27, pp. 62–69, 2010.

go further into the methodology we used. After which the results which could be derived from our experiments are discussed and in the end we will draw some conclusions.

## 2 Research and Methodology

### 2.1 Equipment

The equipment consists of two accelerometer sensors, two data receivers and a data logging unit. The software for monitoring, logging and analyzing of the data is custom made. The MMA7260Q sensors used in the research are tri-axial accelerometers from Freescale. These sensors are mounted on a demo board called the Zstar 1 (represented in Fig. 1) which contains the accelerometer itself, a wireless radio module and an 8-bit MCU which can be re-programmed through a BDM (Background Debug Interface). The radio module creates a wireless link with the 2.4 GHz Zigbee protocol to a USB data receiver plugged in the computer.

Thanks to its wireless characteristics the board allows the sensor to be positioned at many different places on the body. Self-made Velcro straps are used (see Fig. 1) to easily mount the sensors around the upper body and legs. It is important for the reliability of the signal that the sensor cannot move freely and that the straps are as tight as possible. Concerning the study of fall events during the experimental phase of our research, it has to be noticed that every event is recorded on tape. Also the software which analyses the sensor activity is screen captured. Thanks to these recordings every fall event could be studied extensively and because of this possible misdetections could be evaluated more easily.

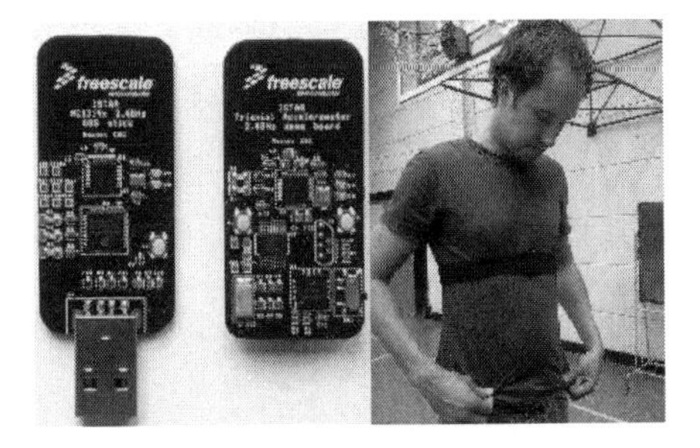


**Fig. 1.** Freescale Zstar sensor board and receiver

### 2.2 Sensor Placement

In order to create a comfortable system for patients and elderly, we wanted to use as less sensors as possible. Whether or not the information drawn from the sensor signal will be useful depends on the position of the sensor on the human body. In the literature a couple of places on the human body that are suitable for our purpose are described. In the end, we ended up with 2 interesting possibilities: a thigh sensor because this results in similar data as a foot-mounted sensor but with smaller signal peak fluctuations and secondly a sensor attached to the torso because this results in very reliable signals thanks to the torso's relatively constant orientation with respect to the user's heading [8]. During the tests we diagnosed that the thigh sensor returns a fluctuating signal because the accelerations are disturbed by the movement of the other leg, the torso sensor on the other hand returns a clean harmonized signal. Because of this we chose to use the torso sensor for our step and fall detection algorithm. Only for our ADL algorithm we used both sensors because by combining the data of these two sensors we we're able to create a basic ADL algorithm that makes it possible to detect the position of the user in real-time. Also during our experiments we found out that the Z-axis data as well as the Y-axis data are sufficient to detect steps, but the Y-axis data show the best recognizable peaks in the signal because the Y-axis represents accelerations up- and downwards. The Z-axis data is less accurate during the process of making turns.

## 2.3 Step Detection

Step detection is the process of determining steps from a dataflow, in our case a dataflow produced by accelerometers. These accelerometers are placed on a person's body to monitor its movements accurately. The basic model of human motion during a walk is a repeated cyclical movement that is remarkably consistent. Yet we can divide this cyclical movement in 2 parts: the so called stance phase when the foot is placed on the ground, and the swing phase when the foot is lifted off the ground and brought forward/backward to begin the next stride [6]. These 2 phases can be recognized in the accelerometer signal where 2 corresponding peaks can be noticed. The first peak occurs when the foot is lifted(swing phase)and the next peak corresponds to the movement in which the heel strikes the ground(stance phase). There are different possibilities available to determine steps from the accelerometer signal. 3 of the most important and widely known algorithms are named by Ying et al [9]:

* Pan-Tompkins Method
* Template-Matching Method
* Peak-Detection method based on combined dual-axial signals

Each algorithm has its own benefits and drawbacks, therefore we've created our own step detection algorithms that tries to combine these several methods into one algorithm that has only minor disadvantages.

**Filters.** For detecting steps we created an algorithm based on filters and peak detection. The filters are used to reduce unwanted high frequency elements and noise in order to properly process the signal data. The first filter is the discrete Kalman filter which reduces the influences of artifacts in the signal. The filter also brings the signals of the three axes at 0g no matter how they are orientated, which makes it easier to use thresholds, detecting zero crossings and calculating min and max values. The second filter is an Average filter to create a smoother signal. It calculates the average value of the last 50 input samples and produces a single output sample. It also removes the high frequency components present in the signal as you can see by

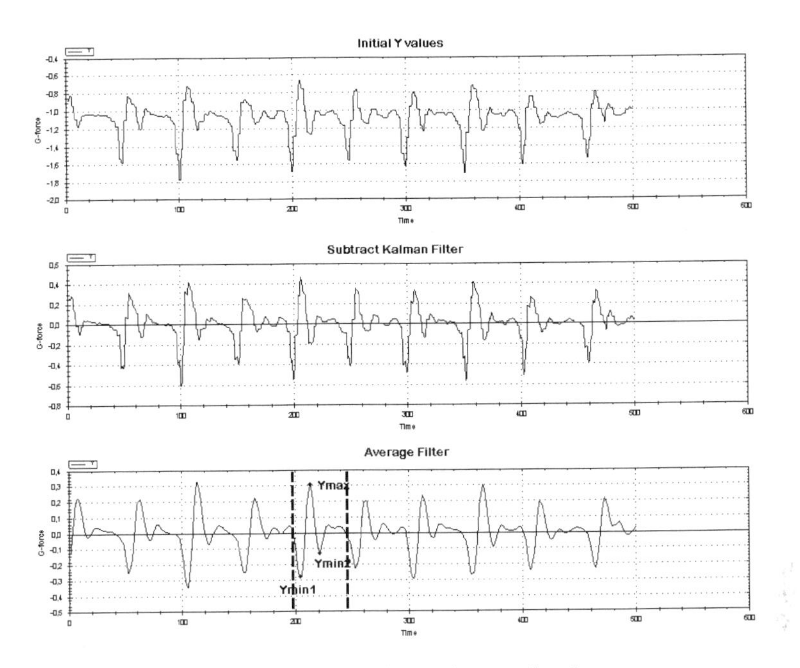


**Fig. 2.** Filters applied to the step signal

observing the sample demonstrated in Fig 2. The result is a clear signal with a recognizable pattern for each step taken.

**Algorithm.** The algorithm that we created is based on a combination of thresholding and peak searching. To easily process the signal we created an array that consists of the last hundred samples of the accelerometer signal. When a new sample is added to the array all other samples shift one place further until the oldest sample is overwritten, also known as the 'first in first out' (FIFO) principle. For every new sample the array is evaluated by the step detection algorithm. If the last one passes a threshold value, the algorithm is triggered. Observe from the third image from Fig. 2 that one step consists of a small negative peak, which occurs when the foot is lifted o_ the ground, followed by a positive peak and again a large negative one when the heel strikes the ground. The peaks vary in amplitude and in time domain. When the last and also largest negative peak is detected by a threshold value, the other (earlier) peaks are already stored in the array and the algorithm can do a checkup. When the next large positive and small negative peaks are detected, the algorithm checks the time span between the two negative peaks. If this time span is small enough, the signal is marked as a step.

**Experiments and Results.** The experiments to validate our step detection algorithm include two parts: an objective validation part and an experimentation part. The experiments are all done on a terrain with various obstacles and different underground. A pedometer was used as a reference. The subjects being monitored were three young adults and two elderly. They were advised to step as naturally as possible.

Part 1: Three human observers monitored the events of the subject and counted its steps taken after which the observers their results we're compared to become inter-observer reliability. In case they obtained a different step count result, the experiment was re-executed. Next to this, another observer who could not see any of the events made a note for every step he could distinguish from the accelerometer signal. These notes were then being compared to the results of the algorithm and the pedometer. The subject did 5 ranges of tests with various step lengths and speed. If we compute the average accuracy of these 5 ranges we should take false positives and negatives into account, resulting into an overall accuracy of 92.40% for the observer. The software on the other hand achieved an overall accuracy of 92,20%. So the pedometer has registered almost the same number of steps as the algorithm did.

Part 2: In our experiment we did several ranges of hundred steps. Some of them included fast walking, other rather slowly or even stumbling. A part of the ranges were done by elderly, others by young adults with or without carrying an object. During our experiments we used the pedometer as a reference. We use the pedometer as a reference because it is a well-known step count device and has a high accuracy although not perfect [7]. The results of our experiment are represented in Table. 1,as you can see the overall accuracy for the pedometer was 97.80% and 97.78% for the algorithm.

### 2.4 Fall Detection

The second part of our research concerns fall detection. Because the end of a fall may be characterized by an impact and horizontal orientation [1], fall detectors have to be able to detect at least one or even better both of these events. Most fall detection systems are based on the shock received by the body when it hits the ground. Detecting the shock can be accomplished by analyzing the sensor data with a threshold technique.

**Different Systems.** As already mentioned before, fall detection is a crucial method to extend the lives of a great group of elderly people. To put this technology into practice, a number of different approaches are proposed.

Fall detection systems can be divided in two main groups: namely the primary fall detection systems that instantaneously detect falls and the secondary fall detection systems which detect falls by the absence of normal activities [1]. These two main groups can again be split up into different subgroups, namely systems based on worn devices and on the other hand Environment sensing systems which require an infrastructure at the patient's location. This final subgroup has major drawbacks such as its cost and intrusiveness. [5]

**Algorithm.** The data from the sensor attached to the torso is as well as it is used to detect steps, now used to detect falls. We know by experiments and research that a fall event is characterized by a large acceleration peak in one or more directions which is followed by a horizontal position. A typical fall event signal is represented in Fig. 3 For the detection of a large acceleration the magnitude vector $r = \sqrt{(x^2 + y^2 + z^2)}$ of the three axes is calculated. If this value passes a certain threshold, the algorithm is triggered. We then wait for the signal to return to a relative normal acceleration. Next, after another small delay of 40 samples the position of the

**Table 1.** Step Detection Statistics

| # | Steps Taken | Pedometer | Algorithm | Type |
|---|---|---|---|---|
| 1 | 100 | 101 | 101 | Normal walking |
| 2 | 102 | 101 | 102 | Normal walking |
| 3 | 101 | 99 | 107 | Normal walking |
| 4 | 99 | 100 | 100 | Normal walking |
| 5 | 100 | 102 | 102 | Normal walking |
| 6 | 100 | 100 | 102 | Fast walking |
| 7 | 100 | 99 | 96 | Stumbling |
| 8 | 100 | 100 | 102 | Normal walking |
| 9 | 100 | 98 | 104 | Doing activities |
| 10 | 100 | 89 | 103 | With charge |
| 11 | 100 | 89 | 101 | With charge |
| 12 | 100 | 100 | 98 | With charge |
| 13 | 100 | 100 | 98 | With charge |
| 14 | 100 | 99 | 102 | With charge |
| 15 | 100 | 98 | 105 | With charge |
| 16 | 100 | 100 | 103 | With charge |
| 17 | 9 | / | 9 | Elder subject |
| 18 | 16 | / | 16 | Elder subject |
|  | 100% | 97.80% | 97.78% |  |

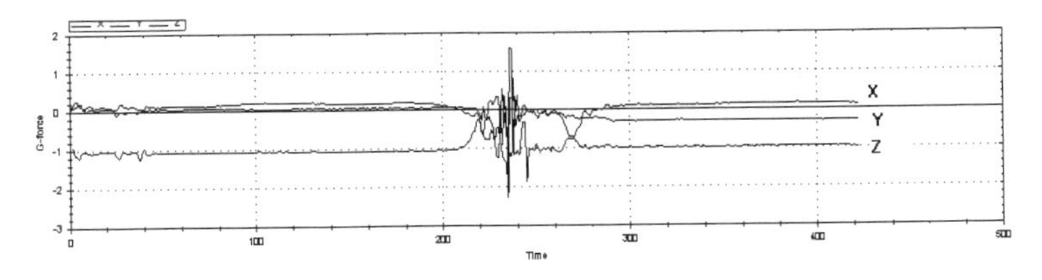


**Fig. 3.** Tri-axial data of a fall event

patient is analyzed. If this turns out to be horizontal, the event is categorized as a fall. To detect the horizontal position, the algorithm checks if the Y-axis is around 0g. Another fall can now not be detected unless the posture of the user has returned to normal and the Y-axis acceleration reverted to -1g. Currently we're working with a sample rate of 20ms which comes down to 50 samples a second In order to detect the large peaks from the impact with the ground we decided not to use filters while analyzing the data for fall events as they may cut off peaks which result in unusable data.

**Experiments and Results** Again the testing is divided in two parts: an objective validation part and an experimentation part.

Part 1: the objective validation was done by one human observer who monitored the events of the subject and noted when a fall had taken place. Another observer which could not see the subject, monitored the signal coming from the acceleration sensor and made a note with a timestamp for every fall event. The notes of these two observatories were then compared with the results of the algorithm. The subject did 6 ranges of tests with a total of 51 events. 35 events were categorized as a fall by the observer. The other events included movements like lying on a bed or bending. The data observer at his turn marked 37 of the events as a fall where the algorithm eventually registered 35 fall events. Two falls were detected as false positive and two as false negative. The results are represented in Table. 2.

Part 2: As we wanted some consistency in the detection schemes we simulated 3 different fall variations: forward falls, backward falls and lateral falls. These different types of falls were carried out by three subjects and were repeated 20 times so we could study 100 falls. Three of these falls were not detected by the algorithm: namely a forward fall with legs straight, one fall with knee flexion and one fall backward with obstruction. This resulted in an error-state of 3% miss-detected falls in this test.

**Table 2.** Fall Detection Statistics

| # | Events | Falls Observed | Data Observer | Algorithm | False Pos. | False Neg. |
|---|---|---|---|---|---|---|
| 1 | 10 | 5 | 6 | 5 | 0 | 0 |
| 2 | 21 | 10 | 12 | 12 | 2 | 0 |
| 3 | 5 | 5 | 5 | 4 | 0 | 1 |
| 4 | 5 | 5 | 5 | 5 | 0 | 0 |
| 5 | 5 | 5 | 4 | 4 | 0 | 1 |
| 6 | 5 | 5 | 5 | 5 | 0 | 0 |
| | | 100% | 93% | 93% | | |

### 2.5 Detecting the Position of the User

So, the placement of two sensors creates the possibility to easily detect three different postures: sitting, lying and standing. These ADL's are detected by properly investigating the signal as done by Culhane et al [2]. The accelerometer sends data from three different axes to the computer. As we mount our sensor one way up, the vertical axis is the Y-axis. In normal position, the Y-axis value is always around -1g. If a user changes position to lying, both sensors at the thigh and torso are oriented horizontally. The Y-axis value of the two sensors is then changed to around 0g as the sensors are orientated horizontally, which can be indicated by using two threshold values. Also a sitting position can be detected. The sensor on the thigh is then orientated horizontally while the one attached to the torso is orientated vertically to the ground. In standing position, both sensors are orientated vertically to the ground.

## 3 Conclusion

The system described in this paper is able to monitor a patient real-time by detecting his posture, count his steps taken and register fall events over a certain time. The posture detection system is very reliable since it properly detects whether the subject is sitting, lying or standing. We call this the detection of the Activities of Daily Living. The tests of the step detection algorithm resulted in values that approached the pedometer results. Nevertheless the pedometer has more false negatives than false positives. This means that the majority of our system's miss-detected steps weren't actually steps. This could be improved by tweaking the algorithm a little by maximizing or minimizing the threshold value and shorten the distance between the negative peaks. As these values are patient dependent we're planning to include automatical thresholding in our future work by which we become a self learning algorithm. Next to this, the fall detection algorithm has also produced very pleasing results. In the test scenario of 5 different falls 97% of the fall events were effectively categorized as such.

## References


1. Bourke, A.K., Scanaill, C.N., Culhane, K.M., O'Brienand, J.V., Lyons, G.M.: An optimum accelerometer configuration and simple algorithm for accurately detecting falls. In: Proceedings of the 24th IASTED International Multi-Conference Biomedical Engineering (2006)
2. Culhane, K.M., Lyons, G.M., Hilton, D., Grace, P.A., Lyons, D.: Long-term mobility monitoring of older adults using accelerometers in a clinical environment. Journal of Clinical Rehabilitation 18, 335–343 (2004)
3. Doughty, K., Lewis, R., McIntosh, A.: The design of a practical and reliable fall detector for community and institutional telecare. Journal of Telemedecine and Telecare 6, 150–154 (2000)
4. Lord, S.R., Sherrington, C., Menz, H.B.: Falls in older people: Risk factors and strategies for prevention. Cambridge University Press, Cambridge (2001)
5. Perolle, G., Fraisse, P., Mavros, M., Etxeberria, I.: Automatic Fall Detection and Activity Monitoring for Elderly. In: Cooperative Research Project - CRAFT
6. Rose, J., Gamble, J.G.: Human Walking. IEEE Press, Los Alamitos (2005)
7. Schneider, P.L., Crouter, S.E.: Bassett, and David, R., Pedometer Measures of Free-Living Physical Activity: Comparison of 13 Models. Official Journal of the American college of sports medicine (2004)
8. Stirling, R., Fyfe, K., Lachapelle, G.: Evaluation of a New Method of Heading Estimation for Pedestrian Dead Reckoning Using Shoe Mounted Sensors. The Journal of Navigation (2005)
9. Ying, H., Silex, C., Schnitzer, A., Leonhardt, S., Schiek, M., Automatic Step Detection in the Accelerometer Signal

# The Costs of Non-training in Chronic Wounds: Estimates through Practice Simulation

Pedro Gaspar[1,a], Josep Monguet[2,b], Jordi Ojeda[2,b], João Costa[1,a], and Rogério Costa[1,a]

[1] Polytechnic Institute of Leiria, [2] Universitat Politècnica de Catalunya
[a] PhD student; [b] PhD
pedrgaspar@gmail.com, jm.monguet@gmail.com,
jordi.ojeda@gmail.com, joao.costa@estm.ipleiria.pt,
rogerio.paulo.costa@gmail.com

**Abstract.** The high prevalence and incidence rates of chronic wounds represent high financial costs for patients, families, health services, and for society in general. Therefore, the proper training of health professionals engaged in the diagnosis and treatment of these wounds can have a very positive impact on the reduction of costs.

As technology advances rapidly, the knowledge acquired at school soon becomes outdated, and only through lifelong learning can skills be constantly updated. Information and Communication Technologies play a decisive role in this field. We have prepared a cost estimate model of Non-Training, using a Simulator (*Web Based System – e-fer*) for the diagnosis and treatment of chronic wounds.

The preliminary results show that the costs involved in the diagnosis and treatment of chronic wounds are markedly higher in health professionals with less specialized training.

**Keywords:** Non-training; pratice simulation; costs estimation.

## 1 Introduction

The prevalence and incidence rates of chronic wounds are substantial in the world [1] and, in addition to the obvious impacts they have on the quality of life, they represent very high economic costs for patients, families, health services, and for society in general [2,3]. For instance, the estimates point to a spending between 2.3 and 3.1 billion pounds (2005-2006 prices) per year in the United Kingdom [4] and 7 billion dollars in the USA, in 2007 (in dressing materials alone), a number that can rise to 20 billion dollars if other costs related to salaries, hospitalization, infection control, among others, are taken into consideration [5].

The training of health care professionals involved in the diagnosis and treatment of chronic wounds plays a decisive role in the control and optimization of costs, simply because by associating the technological advances of materials and equipment with adequate and standard guidelines on prevention and treatment, there will be a very positive impact on the clinical and economic results [6], and, therefore, a large portion of the financial load can be avoided [4].

P. Kostkova (Ed.): eHealth 2009, LNICST 27, pp. 70–75, 2010.

In a context of fast-moving scientific and technological advances, knowledge acquired at school soon becomes outdated [7] and only through *lifelong learning* can we achieve professional updating. Information and Communication Technologies provide the health professionals with the means to access, interpret and apply the organizational knowledge, best practices, competences and skills, leading to more positive clinical results, thus maximizing cost reduction [8].

We have prepared a cost estimate model of *non-training*, using a Simulator (*Web Based System – e-fer*) for the diagnosis and treatment of chronic wounds. The use of such a simulator and of virtual cases has allowed us to overcome methodological problems and ethical constraints we faced when devising the experience, when we searched for empirical evidence to support the general hypothesis that costs vary according to the training of health professionals involved in the diagnosis and treatment of chronic wounds.

## 2 Materials and Methods

**(1)** We developed a *Web Based Learning* (e-fer) that allows us to submit virtual cases of patients with chronic wounds, including the pictorial and non-pictorial information (time of evolution, location, size, tunnelization, oedema, hardening, smell, exudate, pain) on the wound, socio-demographic data, medical history, mobility and nutrition status, and the adequate diagnosis and treatment options. Nine virtual cases of patients with chronic wounds were validated (4 leg ulcers, 3 pressure ulcers, 2 diabetic foot ulcers), through the *double-blind peer review*.

**(2)** We developed a mathematical model to estimate the costs of chronic wounds (Figure1) which, in a society perspective, takes into account variables such as the *Expected Treatment Time, Direct Costs* (dressing costs, labour costs, travel costs from patient home to health service, travel costs from health service to patient home and costs of equipment for pressure relieving support) and *Indirect Costs* (productivity losses by patient, if the patient is a worker, and productivity losses by familiar that support patient. By applying the mathematical model to the virtual cases validated, we prepared the *Optimum Cost* matrices (estimated costs for the optimum treatment options).

**(3)** We developed a *Web Based System* (e-fer Simulator) that allows the simulation of decision-making in the treatment of the virtual cases of chronic wounds. The application of the mathematical model to the options chosen by the users enables the preparation of *Cost of Decision* matrices (estimated costs for the treatment options selected in the e-fer *Simulator*).

**(4)** We conducted a quasi-experimental pre-test/post-test study with a non-equivalent control group, including 12 health professionals in each group. Participants were pre-tested with the e-fer Simulator (nine virtual cases) prior to starting an accredited training programme on the prevention and treatment of chronic wounds (totalling 35 hours of theory and practice). The experimental group alone attended the training programme. Both groups were post-tested with the e-fer Simulator (same nine virtual cases) at the end of training (Figure2). The results were analysed with the Man-Whitney and Wilcoxon nonparametric test, with 95% confidence interval.

$$TC = ETT \times (DCW + LCW + TChs + TCph) + CE + PLF + PLP$$

***Where:***
***TC – Total Costs***
***ETT – Expected Treatment Time (weeks)***

$$ETT = HT(WWIHr + DIHr + ADIHr + APIHr)$$

*HT - Healing Time*
*WWIHr - Wound Washing Impact in healing rate*
*DIHr - Dressing Impact in healing rate*
*ADIHr - Additional Dressing Impact in healing rate*
*APIHr - Additional Procedure Impact in healing rate*

***DCW – Dressing Costs per Week***

$$DCW = DC \times DCP$$

*DCP- Dressing Change periodicity, per week*
*DC- Dressing (materials) Costs, per week*

***LCW –Labour (nursing) Costs per Week***

$$LCW = (LC\min \times 20[1 + 0,25(1 - Smob)] \times DCP)$$

*Smob –Mobility status (1 if the patient is autonomous ; 0 if the patient is not autonomous)*
*LCmin – Labour Costs per minute*

***TChs – Travel costs from patient home to Health Service – per week***

$$TChs = (9,4 \times DCP) \times Smob$$

***TCph – Travel costs from Health Service to patient home – per week***

$$TCph = ([10,6 + 4,4(1 + 0,25[1 - Smob]) + 20LC\min] \times DCP) \times (1 - Smob)$$

***CE – Costs of Equipment for pressure relieving support***

$$CE = PRM + ErgC + PRC$$

*PRM – Pressure relieving Mattress*
*ErgC – Ergonomic Chair*
*PRC – Pressure Relieving Cushion*

***PLF – Productivity losses by familiar that support patient.***

$$PLF = \frac{DPV}{2} \times DCP \times ETT$$

*DPV – Daily production value*

***PLP – Productivity losses by patient (if the patient is a worker)***

$$PLP = DPV \times 7ETT \times Smob \times Age2$$

*Age2 – 0 if ? 65 years old*
*1 if < 65 years old*

**Fig. 1.** Cost Estimation Model

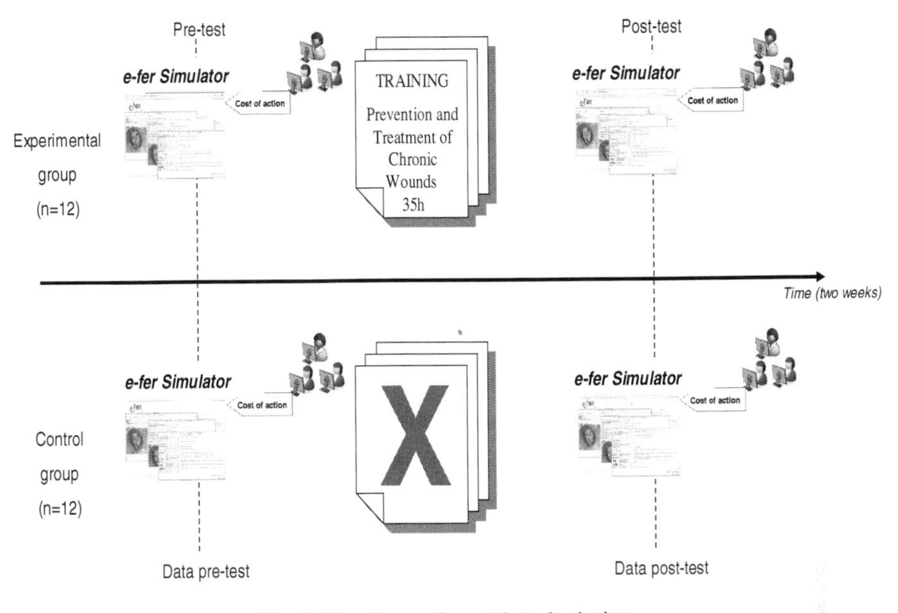


**Fig. 2.** Quasi-experimental study design

## 3 Results

The results reflect the difference between Optimal Costs and the Costs of Action observed in the groups.

The groups were homogeneous with regard to the *Cost of Decision* in the pre-test (U=64; *p*=0.641; 95,5 % CI = -5741,45€ to 4122,89€).

In the Experimental Group, the costs involved in the treatment of virtual cases were fewer in the post-test (figure 3), with a statistically significant difference (Z= -3.059; p= 0.002; 95,8 % CI = -5057,43€ to 13106,51€).

In the Control Group, there were no statistically significant differences (Z=-0.314; p=0.753; 95,8%CI= -3775,78€ to 7836,31€).

Although the Cost of Decision after training (post-test) were more than the *Optimum Costs* in both groups, we noticed that there are obvious differences between the Experimental Group (58.90% above the *Optimum Costs*) and the Control Group (119.19% above the *Optimum Cost*).

If we consider the 2009 prices, the estimated values for the non-training costs total 890.07 € for each wound, and 8,010.60 € for the total treatment of the nine virtual cases. 48% (3,852.73 €) of the savings fall on the indirect costs (salary losses of the reference family member and/or patient) and 52% (4,157.88 €) fall on direct costs. To be more precise, 22% (1,772.1 €) of the savings are related to the nurse's salaries and travel expenses, 19% (1,541.52 €) to the patient's travel expenses to the Health Service, and 11% (873.84 €) to the costs of dressing materials.

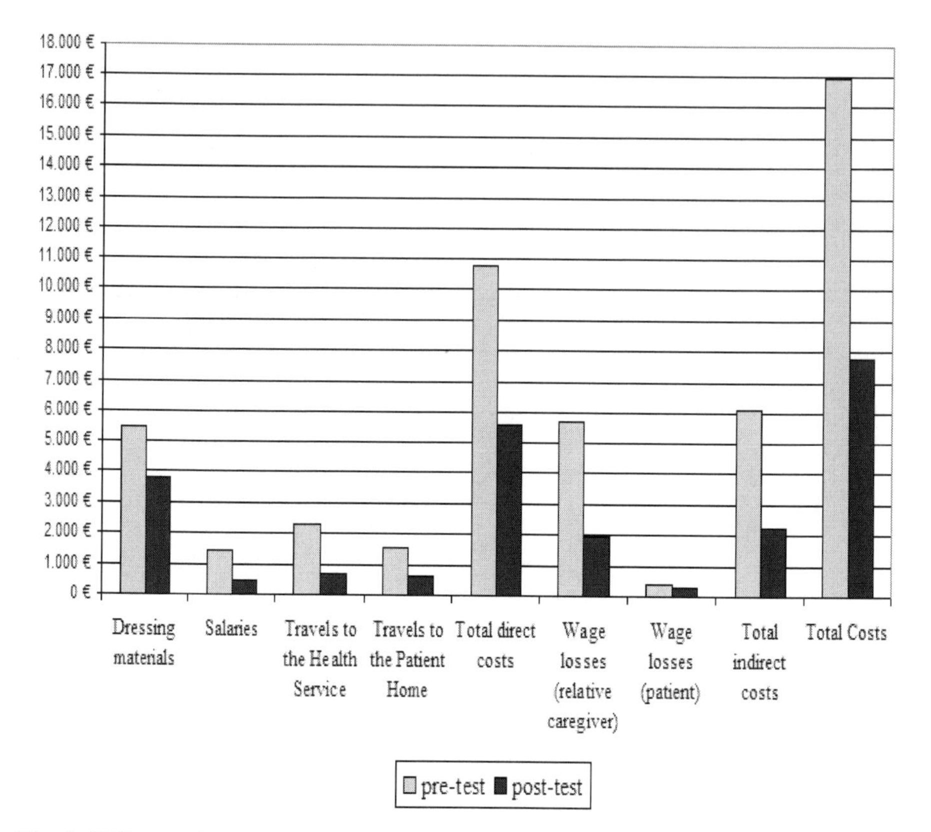


**Fig. 3.** Difference between Optimal Costs and Costs of Decision observed in Experimental Group (before and after training)

## 4 Conclusion

We believe that it is possible to estimate the costs of no training on several areas of health care, and to avoid methodological and ethical constraints in research by following the proposed method. Based on the development of Web Based Learning Systems that allow experts to submit virtual cases built upon real experiences, including all relevant data and the adequate diagnosis and treatment options, this method envisages the development of simulators that allow trainees to diagnose and treat patients in a virtual environment. Finally, by taking on a mathematical model that allows for cost estimation, it is possible to estimate the cost of non-training.

## References

1. Vu, T., Harris, A., Duncan, G., Sussman, G.: Cost-effectiveness of multidisciplinary wound care in nursing homes: a pseudo-randomized pragmatic cluster trial. Fam. Pract., cmm024, 1–8 (2007)

2. Franks, P.J., Bosanquet, N.: Cost-Effectiveness: Seeking Value for Money in Lower Extremity Wound Management. International Journal of Lower Extremity Wounds 3, 87–95 (2004)
3. Janßen, H.J., Becker, R.: Integrated system of chronic wound care healing. EWMA Journal 7, 19–23 (2007)
4. Posnett, J., Franks, P.J.: The burden of chronic wounds in the UK. Nursing Times 104, 44–45 (2008)
5. Zaleski, T.: "Active" Products Drive Wound-Care Market. Goal Is to Offer a Lower Overall Cost Solution (2008), `http://www.genengnews.com/articles/chitem.aspx?aid=2459&chid=0` (accessed May 5, 2009)
6. Dini, V., Bertone, M., Romanelli, M.: Prevention and management of pressure ulcers. Dermatologic Therapy 19, 356–364 (2006)
7. Kokol, P., Blažun, H., Mičetić-Turk, D., Abbott, P.A.: e-Learning in Nursing Education – Challenges and Opportunities. In: Park, et al. (eds.) Consumer-Centered Computer-Supported Care for Healthy People. IOS Press, Amsterdam (2006)
8. Nelson, A.: E-learning. A Pratical Solution for Training and Tracking in Patient-care Settings. Nursing Admtnistration Quarterly 27(1), 29–32 (2003)

# Open Source Virtual Worlds and Low Cost Sensors for Physical Rehab of Patients with Chronic Diseases

Salvador J. Romero[1,2], Luis Fernandez-Luque [2], José L. Sevillano[1], and Lars Vognild[2]

[1] Robotics & Computer Technology for Rehabilitation Laboratory, University of Sevilla, Spain
[2] Northern Research Institute, Tromsø, Norway
{luis.luque,lars.vognild}@norut.no

**Abstract.** For patients with chronic diseases, exercise is a key part of rehab to deal better with their illness. Some of them do rehabilitation at home with telemedicine systems. However, keeping to their exercising program is challenging and many abandon the rehabilitation. We postulate that information technologies for socializing and serious games can encourage patients to keep doing physical exercise and rehab. In this paper we present Virtual Valley, a low cost telemedicine system for home exercising, based on open source virtual worlds and utilizing popular low cost motion controllers (e.g. Wii Remote) and medical sensors. Virtual Valley allows patient to socialize, learn, and play group based serious games while exercising.

**Keywords:** Tele-rehabilitation, Virtual Worlds, Open Source, Low-Cost.

## 1 Introduction

Most patients with chronic diseases depend on exercising and rehab to stabilize the progression of their disease and increase quality of life [1]. However, many of them suffer motivation problems which could lead into poor compliance to perform the physical exercises suggested by their healthcare professionals. Patients become more sensible to depression when they are no longer able to go out from their homes. They cannot go to the health center for rehabilitation anymore, and have to exercise at home. They stop having a sense of group belonging and tend to feel isolated.

Many of the solutions proposed or adopted to confront this problem are based on computer applications. Usually, a system based on a PC and medical sensors is installed in the patient's house, to help doctors monitor and supervise the patient´s rehabilitation [2]. For the exercises, systems use to have some hardware, like intelligent clothes or a bike [3]. They also provide communication tools, such as videoconference.

In recent years, virtual worlds like Linden Research's Second Life have become really popular. They have shown to be a great tool for collaborative works, learning and e-health [4-6]. An increasing number of companies, universities, and educational centers have presence in Second Life. Users can attend virtual lectures, interact with applications and 3D objects, and socialize with other users. They can see if other people are around and can start a conversation with them. These features make virtual worlds a natural and attractive social platform.

P. Kostkova (Ed.): eHealth 2009, LNICST 27, pp. 76–79, 2010.

## 2 Virtual Valley

We are taking advantage on innovations in virtual world technologies, game-based low-cost solutions for motion tracking [7], and medical sensors in the development of our approach; a social-oriented virtual world controlled by sensors, where the patient can learn, exercise, and interact with others from home.

Although Second life is a good option to reach as many people as possible, due to sensible data we manage and requirements of restrict access, we need to deploy the virtual world in our own servers. For that, there are open source clones and alternatives to Second Life, e.g. OpenSimulator and Sun Wonderland [8]. We have decided to use Wonderland, as we explain in the following section.

### 2.1 System Architecture

Virtual Valley implements a client-server architecture, running a Glashfish server on the server side, the Wonderland server and other complementary software (Fig.1).

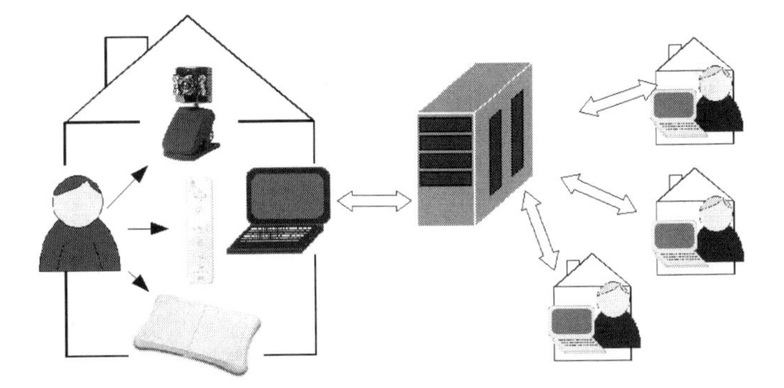

**Fig. 1.** System architecture

Wonderland is a Java open source software for virtual worlds. It is able to run Java and X11 compatible applications embedded inside the virtual world, which users can interact with through their avatars. This feature provides a straightforward way to develop custom multiuser applications inside the Virtual World, and makes Wonderland very flexible. Another interesting feature of Wonderland is the voice bridge technology, which allows the user to have a voice chat at any moment. The user just speaks through her microphone, and nearby avatars can hear her.

On the client side, a PC is running the Wonderland client, controlled and receiving input from sensors, such as the Wii Remote controller and the Nonin Pulsioximeter. The communication protocols of these Bluetooth sensors are publically available. There is a growing set of libraries and APIs[1] which allow applications to interact with them easily. We are using WiiGee [9] for gesture recognition with the Wii Remote and Bluecove to enable Bluetooth communications.

Thanks to the voice bridge technology and the use of motion controllers and sensors, a keyboard and mouse are not necessary.

---

[1] *Windows*: GlovePie, *OSX:* Darwiin remote; *Linux*: Cwiid driver; *Multiplatform*: Wiiuse (C), Wiimotelib (.Net); MoteJ, Wiimote Simple, WiimoteJ, Wiigee (Java), WiiFlash (Flash!).

### 2.2 Functional Design

The virtual world shown to the user is a 3D representation of a Valley, where several buildings can be found: 1) the Learning Center, 2) the Virtual Gym, for exercising; and 3) the amusement garden, which is basically a game oriented virtual gym.

The *Learning Center* building is intended to aid patients to learn how to deal with their illness. The info is shown on walls, like in a museum. It can be slides with text and images, videos, websites and interactive applications. The learning center has a conference room, where a person controlling an avatar can give a talk. By moving her avatar after a microphone, she can be heard in the whole room. The speaker is able to show slides, applications or live videos on a couple of blackboards nearby.

The *Virtual Gym* building (Fig. 2) is dedicated to exercise and rehabilitation. Inside the building there are 3D items, called animatics, which resemble a treadmill, with a large screen in front of it. When an avatar steps on the animatic, the view is centered on the animatic screen. There is then displayed the exercising program, adaptable to each user's needs, that the user has to perform using the sensors. There is another conference room used for group exercising leaded by a physiotherapist.

Avatars perform predefined movements, the same in all cases, to show that a user is exercising. This might help patients to get over their shames and socialize with others, including healty users (e.g. relatives).

**Fig. 2.** Screen captures of a prototype. From left to right, Virtual Gym and Music Hall.

The *Amusement Garden* is similar to the virtual gym, but oriented to games and group activities. The technique of disguising rehabilitation exercising with serious games has been successfully tried in several occasions, e.g. [10]. The avatar will use a modified animatic to fit the context of the game. Several of these activities are: 1)*Garden scout*: The avatar rides an animatic which resembles a bike, a scooter or similar. As she performs her exercises, the animatic-bike moves around the garden; 2) *Sailing the Fjord:* Similar to the previous one, but group oriented; 3) *Music hall* (Fig. 2). Avatar takes places in an orchestra of animatic-instrument. When they exercise properly, their instruments sound.

## 3 Discussion and Future Work

Virtual Valley represents the convergence of traditional telerehabilitation and new ways of communication and social interaction. Based on previous experiences [2], our hypothesis is that this kind of social-oriented approaches can help encouraging and motivating patients to exercise and keep doing rehabilitation.

We have done a study of the art and tested the most promising technologies. The Virtual Valley System has been designed with the active implication of healthcare professionals of the University Hospital of North Norway, Tromsø The system is currently being implemented, and we estimate that a working version for evaluation by users will be ready by the third quarter of the present year.

Thanks to the use of state of the art low cost sensors and open source software, we are in condition to build an affordable system that we can install in most patients' houses. The open source community is active in these areas, and the success that low cost game and medical sensors are achieving in the market, guarantee a continuous development of software, hardware and APIs which will support our project. The virtual world can be expanded easily, and new applications, activities and buildings can be added as needed. Other sensors can be used and more function added to those we already use, like precise 3D tracking.

**Acknowledgments.** This project is a part of the MyHealthService project from the Tromso Telemedicine Laboratory, co-funded by the Research Council of Norway. The project has been partially supported by the Fidetia and the Telefonica Chair on Intelligence in Networks from the University of Sevilla (Spain).

## References

1. W. H. Organization: Global Strategy on Diet, Physical Activity and Health, Facts Related to Chronic Diseases, `http://www.who.int/dietphysicalactivity/publications/facts/chronic/en/index.html` (Cited: 4 7, 2009)
2. Burkow, T.M., Vognild, L.K., Krogstad, T., Borch, N., Ostengen, G., Bratvold, A., Risberg, M.J.: An easy to use and affordable home-based personal eHealth system for chronic disease management based on free open source software. Studies in health technology and informatics, vol. 136 (2008)
3. Wang, Z., Kiryu, T., Tamura, N.: Personal customizing exercise with a wearable measurement and control unit. Journal of NeuroEngineering and Rehabilitation 2(1), 14 (2005)
4. Hansen, M.M.: Versatile, Immersive, Creative and Dynamic Virtual 3-D Healthcare Learning Environments: A Review of the Literature. EdD, MSN, RN: JMIR (2008)
5. Gorini, A., et al.: A Second Life for eHealth: Prospects for the Use of 3-D Virtual Worlds in Clinical Psychology. JMIR (2008)
6. Galego, B., Simone, L.: Leveraging online virtual worlds for upper extremity Rehabilitation. In: IEEE 33rd Annual Northeast Bioengineering Conference, NEBC 2007 (2007)
7. Smith, B.K.: Physical Fitness in Virtual Worlds. Computer 38(10), 101–103 (2005)
8. Project Wonderland, `https://lg3d-wonderland.dev.java.net`
9. WiiGee: A Java-based gesture recognition library for the Wii remote, `http://www.wiigee.org`
10. Consolvo, S., Klasnja, P., Mcdonald, D.W., Avrahami, D., Froehlich, J., Legrand, L., Libby, R., Mosher, K., Landay, J.A.: Flowers or a robot army?: encouraging awareness & activity with personal, mobile displays. In: UbiComp 2008: Proceedings of the 10th international conference on Ubiquitous computing, pp. 54–63 (2008)

# Data Triangulation in a User Evaluation of the Sealife Semantic Web Browsers

Helen Oliver, Patty Kostkova, and Ed de Quincey

City eHealth Research Centre (CeRC), City University London,
Northampton Square, London, UK
{helen.oliver.1,ed.de.quincey}@city.ac.uk, patty@soi.city.ac.uk

**Abstract.** There is a need for greater attention to triangulation of data in user-centred evaluation of Semantic Web Browsers. This paper discusses triangulation of data gathered during development of a novel framework for user-centred evaluation of Semantic Web Browsers. The data was triangulated from three sources: quantitative data from web server logs and questionnaire results, and qualitative data from semi-structured interviews. This paper shows how triangulation was essential in validation and completeness of the results, and was indispensable in ensuring accurate interpretation of the results in determining user satisfaction.

**Keywords:** Semantic Web Browsers, User Evaluation, Data Triangulation, Healthcare Ontologies, Sealife.

## 1 Introduction

The Semantic Web (SW), as a realisation-in-progress of the original vision of the World Wide Web [1], aims to increase findability of specific information among the many results returned by a Web search. Semantic Web Browsers (SWBs) are emerging as a potential solution, but little attention has been paid to evaluating these browsers to assess real-world user satisfaction.

In the course of the EU-funded Sealife project [2], we addressed this lack by developing an innovative framework for user-centred evaluation of Semantic Web Browsers [3] for the life sciences using data from 3 sources: web server logs, questionnaire results, and semi-structured interviews. The data provided invaluable insight into user thought processes and satisfaction with the SWBs.

However, it is essential to bring together the quantitative and qualitative results in order to draw the appropriate conclusions about user satisfaction. In this paper we discuss an adaptation of a triangulation method, and the triangulated results of the Sealife SWB evaluation, demonstrating the necessity of data triangulation to ensure accurate interpretation of the collected data. We show how the impression received from one type of data can be dramatically altered by another type of data.

P. Kostkova (Ed.): eHealth 2009, LNICST 27, pp. 80–87, 2010.

## 2 Background

Most evaluations of web portals combine qualitative data (e.g. from interviews and focus groups) with quantitative data (from weblog analysis, standard questionnaires, etc.). As each source accumulates data in answer to a particular question, combining data sets is essential to paint a more complete picture of user acceptance. Triangulation has been investigated in evaluations of the impact of digital libraries (DLs) of medicine on clinical practice [4]; of electronic transmission of medical findings [5]; and of nursing documentation systems [6]. Given the ever-increasing interest in Semantic Web (SW) technologies in the life sciences, user-centred evaluation making use of triangulation is indispensable in producing much-needed results.

Not only has little attention thus far been paid to user-centred evaluation of SWBs, Ammenwerth [6][7] has pinpointed a lack of attention to data triangulation as a major weakness in user evaluations of health information systems. Despite substantial contributions [4][5][6] this need still has not been addressed in such comparable user evaluations as have been done on SWBs [8][9]. We investigated triangulation methodology in our evaluation of SWBs for the Sealife project.

### 2.1 Sealife SWB Evaluation

The framework we created for the Sealife evaluation was tested in the first user-centred evaluation of SWBs of its kind. It was a within-subjects [10] evaluation of three SWBs for the life sciences, using live, real-world systems with established user bases as control platforms, and recruiting study participants from the real-world target audiences of the SWBs.

The SWBs that were evaluated were the three Sealife browsers: COHSE [11], a CORESE-based SWB [12], and GoPubMed [13] and its related system GoGene as well as an extended version of GoPubMed. The control platform for COHSE and the CORESE-based SWB was the NeLI Digital Library (DL) [14], which has infectious disease professionals as its target audience. The control platform for GoPubMed/GoGene, which have microbiologists as their target audience, was PubMed [15].

A detailed breakdown of methods and results is beyond the scope of this paper, but more information can be found in [3]. This paper will focus on the aspects of the study relevant to triangulation.

The evaluations were presented in web format and began with a pre-questionnaire regarding demographics and previous experience of the control platform. There followed a number of information-seeking tasks tailored to the SWB. After each task was a post-task questionnaire with two questions regarding findability and ease of use, and the evaluation ended with a post-questionnaire asking for users‘ ratings of both the control platform and the intervention SWB. The evaluation was conducted both online and in workshops; workshop participants were asked to give semi-structured interviews.

## 3 Use of Triangulation for Semantic Web

The triangulation in this study combined qualitative and quantitative data from three sources [16].

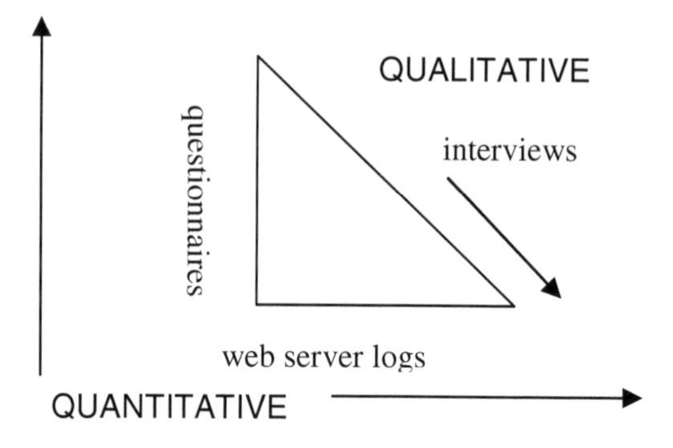


**Fig. 1.** Triangulation of data: qualitative and quantitative dimensions

The quantitative data was provided by the web server logs and questionnaires eliciting ratings of both the control and intervention systems; the qualitative data, by the semi-structured interviews, conducted during workshops with those participants who had the time to be interviewed. Because not all of the participants were interviewed, this evaluation was not fully triangulated; further study with full triangulation or sampling is needed and we envisage future work to develop a framework for triangulation of data in user-centred evaluation of SWBs.

### 3.1 Value of Data Triangulation in Interpreting the Results

The web server logs provided measures of time taken by each user to complete the tasks, and of usage of semantic links against non-semantic links, number of external pages viewed; and whether or not users viewed the target documents containing the answers to the tasks. The post-task and post-evaluation questionnaires gathered users' ratings of both the control platform and the intervention SWB in terms of information *findability, usability,* overall system *speed, relevance* of results, and overall *likeability* of the system. However, questionnaires could only elicit answers to the questions the evaluators thought to ask; the semi-structured interviews were essential for completeness [17] of results; in particular, observation of workshop users would tell us how intuitive they found the SWBs, complementing the questionnaire responses regarding usability. In the next section, we will illustrate how triangulation's core benefits of validation and completeness were demonstrated in our study. [6][17]

## 4 Sealife Results

The validity and completeness of the Sealife results were attained with triangulation of the web server log and questionnaire data with semi-structured interviews of some of the participants. COHSE was evaluated by 67 participants, 39 online and 28 in workshops. The CORESE-based SWB was evaluated by 14 participants; 2 online (only one of whom completed the evaluation) and 12 in workshops. GoPubMed was evaluated by 137 participants online and 4 in an informal workshop setting where full interviews were not conducted. GoGene and an extended version of GoPubMed were evaluated by 14 participants in a single workshop. The triangulated results are not statistically significant because of the low numbers of interviewees per workshop, and so although some of our triangulated data is quantitative, our *interpretation* of the triangulated data is entirely qualitative. In this section, we will discuss the data gathered from the three sources of logs, questionnaires, and interviews in order to demonstrate that it is indispensable to combine them by triangulation.

### 4.1 Web Server Logs

Behind the gathering of data were a number of implicit assumptions. In considering time taken to complete tasks, we assumed that faster completion of tasks was better and that greater use of semantic links was better, where "better" equates to greater user satisfaction. The assumptions become risky if applied uncritically to a single dimension of the data. Ammenwerth has explained the thinking behind the quantitative approach thus: "The results of a measurement are clearly interpretable. Any subjective interpretation is not helpful, and therefore, has to be avoided." [6] On the other hand, Brown [18] argues: "Anyone expecting to arrive at a picture of user-behaviour from web-log analysis is likely to be disappointed." Table 1 shows the average times spent using each system, gathered from the web server logs:

**Table 1.** Average time for all tasks by all users on each system in seconds

| GoGene | GoPubMed | COHSE | CORESE | PubMed | NeLI |
|---|---|---|---|---|---|
| 229 | 126 | 478 | 266 | 194 | 387 |

This shows that tasks completed using PubMed were completed more quickly than the GoGene tasks. If, as we assumed, faster is better, did users prefer the control platform of PubMed to the slower, implicitly "worse" intervention SWB of GoPubMed? The logs only tell us the speed at which users worked; it does not tell us how that speed affected, or was affected by, user satisfaction. The significance of the web log data could only be determined by asking the users.

### 4.2 Questionnaires

Details of the questionnaires can be found in [3]; the results paint an informative picture of users' attitudes. GoPubMed/GoGene were rated the highest in the dimensions of likeability, information findability, relevance, and system speed. The one dimension in which GoPubMed/GoGene did not "win" was usability; the questionnaires seemed to portray COHSE as the most usable of the SWBs. Overall, the GoPubMed and GoGene semantic browsers received far more positive ratings than either COHSE or the CORESE-based SWB, with more and larger differences in mode ratings between the control system (PubMed) and the intervention system. In no case did GoPubMed/GoGene receive *worse* mode scores than the control platform, whereas COHSE and the CORESE-based SWB received several inferior mode scores.

### 4.3 Semi-structured Interviews

Examining the questionnaire results alone might lead us to believe that GoPubMed/GoGene, the overall "winner", was reasonably well liked but that COHSE was considered the most usable. However, questionnaire respondents could only answer the questions that the evaluators chose to ask. GoPubMed included free text fields, which elicited important feedback about accessibility: "Looks great be careful with the colors as dyslexic people find some color difficult to read". It was in the workshops, however, that the most dramatic discrepancies between our assumptions and reality were revealed. It became apparent early that user interface (UI) maturity was a fundamental, rather than a superficial, concern, with the unpolished UIs of the university-developed research applications COHSE and CORESE a serious impediment to usability. The maturity of the much more abundantly-resourced GoPubMed/GoGene UI elicited critiques at a much higher level of functionality than the other SWBs, which were difficult for participants to use at all. To test intuitiveness, the early workshops presented the SWBs with minimal introduction. Online evaluations had been running for some time, but observing user behaviour, and hearing interview feedback, immediately made it clear that the SWBs were not intuitive to their target audiences as they were to us as computing professionals. Later workshops were preceded by brief explanatory presentations, which reduced users' confusion but were not (we were told) in-depth enough to eliminate it. Most startlingly, it was discovered at the earliest workshops that many users could not tell the difference between the control platform and the intervention SWB, and much of the feedback from the first set of interviews turned out to be critiques of the control platform, the NeLI DL. When one such user was asked her opinion of the COHSE link boxes, the participant replied: "Those awful little boxes? They were really distracting, I didn't really understand what they were." Explanatory presentations eliminated the problem, but users still expressed difficulty: one cited the "busy-ness" of the NeLI home page as a source of confusion between the control and intervention platforms, and another remarked that there was "not much difference" between the NeLI DL alone and the NeLI DL enhanced by the COHSE service.

## 5 Sealife Evaluation: Validation and Completeness of Results

The value of triangulation is that it provides *validation* and *completeness* to the results of a study. [17][6]. This was certainly the case with the Sealife evaluation.

### 5.1 Validation

We were somewhat expecting triangulation of user data to show discrepancy [17] between what users said and what they did, and between statements made in person and responses entered into web forms. This was certainly the case for COHSE's findability ratings – at workshops where some users rated this as adequate or good, the logs showed that none of that session's participants had actually found the answer, (which was very specific and contained in a single target document). Other than this, we found that individuals who were interviewed tended to be consistent in their interviews, questionnaire responses, and logged actions. One user worked quickly through the COHSE tasks and was so unusually positive in her ratings and comments about it that we suspected her responses were not genuine and should be discounted. However, the weblogs showed that time spent on each task was between one and two minutes per task: fast, but two others were faster. Logs also showed that she activated

4 link boxes, which matched the median number for all respondents. She viewed only one external page, which one might seize upon as confirmation of duplicity, only to realise that some users did not visit any external pages, and among those who did, one page was the mode.

### 5.2 Completeness

Interestingly, while COHSE interviewees who rated the SWB negatively often had spent substantial time on each task (more than the expected 5 minutes, and more than they spent on the control platform), several GoPubMed/GoGene users who spent more time on GoGene than on PubMed or the extended GoPubMed spoke of GoGene as their favourite and rated it highly in the questionnaires. One respondent spent just under 14 minutes on the four PubMed tasks, just under 10 minutes on the three extended GoPubMed tasks, and just under 19 minutes on the four GoGene tasks. She gave the PubMed tasks a high rating (92% of the maximum score), the extended GoPubMed tasks 67% of the maximum score, but GoGene 100%. She stated in the interview that the SWBs were "very useful tools" but also mentioned difficulties using the extended GoPubMed. Two other users showed similar patterns in their triangulated data, spending the longest time on the GoGene tasks but rating and describing it as the best one. We therefore cannot jump to the conclusion that spending more time completing tasks implies that the SWB is worse (or better).

## 6 Discussion

The GoPubMed/GoGene workshop tended to confirm positive impressions; the CORESE-based SWB workshop confirmed the negative questionnaire results. However, the GoPubMed/GoGene workshop also confirmed that the issues with this SWB were the most trivial and that the *somewhat* higher questionnaire ratings mask a *dramatically* better user experience. While the other SWBs were rated rather negatively by the questionnaires, impressions of COHSE's greater usability were quashed by contact with the users in person; and the severity of users' problems would have gone undetected without interviews. We had hoped to gather observational data of user actions in situ, and the use of eye tracking software was considered, but time constraints prevented implementation of this or other forms of recording such as video; this will inform planning for future work. While the study fell short of complete triangulation because not all participants were interviewed, recruitment of in-person participants, particularly of busy clinicians, was difficult and resource-intensive. Recruiting enough to attain statistical significance for all data sources would have been impractical even had it been possible. In future work, careful sampling of a subset of individuals for interview might be a better solution than trying to interview 100% of a large number of participants.

## 7 Conclusion

We have developed a method of triangulating quantitative and qualitative data in user centred evaluation of SWBs, addressing a need for greater attention to a technique which is essential for accurate interpretation of data. Having previously applied the framework we developed for user-centred evaluation of SWBs, we triangulated quantitative data from the web server logs and from questionnaires eliciting ratings of users' satisfaction with a number of dimensions of the system, with qualitative data from semi-structured interviews eliciting users' opinions on matters which were important to the users but which had not necessarily been considered by the evaluators. This triangulation was demonstrated to be essential in building up a true interpretation of the results, as impressions built up from one type of data changed dramatically in light of another type of data. Answers about system speed were provided by log data, but the meaning of the answers could only be found in the questionnaires and interviews. Questions about usage of semantic links compared with non-semantic links, and whether or not users found the answers to tasks, could only be answered by log data; but questionnaires and interviews revealed discrepancies between users' reports and their actions. Questions about the intuitiveness of the system were partly answered by questionnaire results, but the full meaning and significance of the results was only discovered in the interviews.

The ultimate question about user satisfaction was only answerable by triangulating the data from all three sources. If any one of the three modes of data collection had been excluded, the evaluation results might have been severely misinterpreted.

**Acknowledgments.** This paper is a direct result of the work of Gawesh Jawaheer, Gemma Madle (CeRC); Dimitra Alexopoulou, Michael Schroeder (TU Dresden); Bianca Habermann (Scionics); Simon Jupp, Robert Stevens (University of Manchester); and Khaled Khelif (INRIA Sophia-Antipolis).

## References

1. Berners-Lee, T., Hendler, T., Lassila, O.: The Semantic Web. Scientific American 284, 34–43 (2001)
2. Schroeder, M., Burger, A., Kostkova, P., Stevens, R., Habermann, B., Dieng-Kuntz, R.: From a Service-based eScience Infrastructure to a Semantic Web for the Life Sciences: The SeaLife Project. In: Workshop on network tools and applications in biology, NETTAB 2006, Pula, Italy (2006)
3. Oliver, H., Diallo, G., de Quincey, E., Alexopoulou, D., Habermann, B., Kostkova, P., Schroeder, M., Jupp, S., Khelif, K., Stevens, R., Jawaheer, G., Madle, G.: A User-Centred Evaluation Framework For The Sealife Semantic Web Browsers. BMC Bioinformatics, Special Issue (in press, 2009)
4. Madle, G., Kostkova, P., Mani-Saada, J., Roy, A.: Lessons Learned From Evaluation of The Use of The National Electronic Library of Infection. Health Informatics Journal 12, 137–151 (2006)
5. Machan, C., Ammenwerth, E., Schabetsberger, T.: Evaluation of the Electronic Transmission of Medical Findings from Hospitals to Practitioners by Triangulation. Methods of Information in Medicine 44, 225–233 (2005)
6. Ammenwerth, E., Iller, C., Mansmann, U.: Can evaluation studies benefit from triangulation? A case study. International Journal of Medical Informatics 70, 237–248 (2003)

7. Ammenwerth, E., Gräber, S., Bürkle, T., Iller, C.: Evaluation of Health Information Systems: Challenges and Approaches. In: Spil, T.A.M., Schuring, R.W. (eds.) E-health systems diffusion and use, pp. 212–236. IGI (2005)
8. Reichert, M., Linckels, S., Meinel, C., Engel, T.: Student's Perception of A Semantic Search Engine. In: Proceedings of the IADIS Cognition And Exploratory Learning In Digital Age (CELDA 2005), Porto, Portugal, pp. 139–147 (2005)
9. Uren, V., Motta, E., Dzbor, M., Cimiano, P.: Browsing For Information By Highlighting Automatically Generated Annotations: A User Study And Evaluation. In: K-CAP 2005: Proceedings of the 3rd International Conference on Knowledge Capture, pp. 75–82. ACM Press, New York (2005)
10. Hoeber, O., Yang, X.D.: User-Oriented Evaluation Methods for Interactive Web Search Interfaces. In: 2007 IEEE/WIC/ACM International Conferences on Web Intelligence and Intelligent Agent Technology Workshops, pp. 239–243. IEEE CS Press, Los Alamitos (2007)
11. Yesilada, Y., Bechhofer, S., Horan, B.: Dynamic Linking of Web Resources: Customisation and Personalisation. In: Wallace, M., Angelides, M.C., Mylonas, P. (eds.) Advances in Semantic Media Adaptation And Personalization. Springer Series on Studies in Computational Intelligence, vol. 93, pp. 1–24. Springer, Berlin (2008)
12. Diallo, G., Khelif, K., Corby, O., Kostkova, P., Madle, G.: Semantic Browsing of a Domain Specific Resources: The Corese-NeLI Framework. In: IEEE/WIC/ACM International Conference on Web Intelligence and Intelligent Agent technology, WI-IAT 2008, pp. 50–54. IEEE, Sydney (2008)
13. Doms, A., Schroeder, M.: GoPubMed: Exploring Pub. Med. with the Gene. Ontology. Nucleic Acids Research 33 (Web Server Issue), W783–W786 (2005)
14. Diallo, G., Kostkova, P., Jawaheer, G., Jupp, S., Stevens, R.: Process of Building a Vocabulary for the Infection Domain. In: 21st IEEE International Symposium on Computer-Based Medical Systems, pp. 308–313. IEEE, Los Alamitos (2008)
15. Pub. Med., `http://www.ncbi.nlm.nih.gov/pubmed/`
16. Denzin, N.: Strategies of Multiple Triangulation. In: Denzin, N. (ed.) The Research Act. a theoretical introduction for sociological methods, pp. 297–331. Aldine, Chicago (1970)
17. Greene, J., McClintock, C.: Triangulation in Evaluation. Evaluation Review 9, 523–545 (2005)
18. Brown, S., Ross, R., Gerrard, D., Greengrass, M., Bryson, J.: RePAH: A User Requirements Analysis for Portals in the Arts and Humanities - Final Report. Arts & Humanities Research Council, UK (2006)

# Adaptive Planning of Staffing Levels in Health Care Organisations

Harini Kulatunga[1], W.J. Knottenbelt[1], and V. Kadirkamanathan[2]

[1] Department of Computing, Imperial College London,
180 Queen's Gate, London SW7 2AZ, United Kingdom
{hkulatun,wjk}@doc.ic.ac.uk

[2] Department of Automatic Control and Systems Engineering,
University of Sheffield,
Mappin Street, Sheffield S1 3JD, United Kingdom
visakan@sheffield.ac.uk

**Abstract.** This paper presents a new technique to adaptively measure the current performance levels of a health system and based on these decide on optimal resource allocation strategies. Here we address the specific problem of staff scheduling in real-time in order to improve patient satisfaction by dynamically predicting and controlling waiting times by adjusting staffing levels. We consider the cost of operation (which comprises staff cost and penalties for patients waiting in the system) and aim to simultaneously minimise the accumulated cost over a finite time period. A considerable body of research has shown the usefulness of queueing theory in modelling processes and resources in real-world health care situations. This paper will develop a simple queueing model of patients arriving at an Accident and Emergency unit and show how this technique provides a dynamic staff scheduling strategy that optimises the cost of operating the facility.

**Keywords:** Adaptive staff scheduling, staffing cost minimisation, integrated health care systems.

## 1 Introduction

Health care systems are gradually evolving from disparate general practices and hospitals to integrated care delivery systems [1]. In this context developing system-wide integration of administration, clinical care, information technology (IT), and financing is the ultimate goal. It has been found that highly centralized networks had better financial performance than did those in more decentralized networks [2]. One of the important operational issues in health care involves resource planning such that the goals of high resource utilisation, meeting patient response times and minimising cost are met [3]. A general modelling and solution methodology found in the existing literature is a *steady-state* queueing network model and an optimisation framework to guide resource planning decisions [5], [4]. The authors believe that most methods are more appropriate either when

P. Kostkova (Ed.): eHealth 2009, LNICST 27, pp. 88–95, 2010.

patient arrival rates are expected to remain relatively constant or when all possible uncertainties in patient arrivals are known in advance so that an allocation policy for a fixed time period can be predetermined. However in an Integrated Health Care (IHC) system such fixed optimisation strategies are not the most efficient since total knowledge of all possible scenarios will be impossible. Such a policy for all times can lead to insufficient/wasted resources over some time periods and ultimately increase in cost.

For management of a IHC system, queueing models can provide an analysis capability which can improve the timeliness of interventions and be helpful in the process of external scrutiny (e.g. by the Health Care Commission). Furthermore they can be a starting point for verifying the credibility of reported performance by different IT system solutions. This paper presents a solution methodology based on *transient* queueing models as opposed to steady-state models which can only provide information about the real-world system under the assumption that the system is stationary. On the other hand a transient queueing model is a more robust approach in providing approximate real-time information about a health care system. In this paper the queueing model is combined with a dynamic optimisation technique which determines a resource allocation policy based on instantaneous system performance measurements and guarantees a minimised accumulated cost over a finite time period. To be specific, the model of the health care system *adaptively* determines the optimal staffing level (i.e. number of clinicians) required to achieve a target level of customer service (i.e. target waiting times) and minimises the staffing related cost of operating the facility. In essence this is an attempt to answer the following question,

At a given time which resource allocation strategy gives the optimal cost?

and provide a financial planning model for IHC administrators and financial managers to study and evaluate the economic impact of changes in a organization's resources at a given time.

The rest of the paper is organised as follows. The first section explains the general staffing model considered in this paper. Next section which describes the solution method for the model, presents the relation between accumulated cost, staffing level and waiting times of patients. The section also describes the adaptive technique to determine the optimal staffing policy over a finite time duration. This is followed by a section on how this general model relates to the specific problem of an Accident and Emergency (A&E) department. Finally simulation results are generated based on actual input data gathered from an NHS centre in London. This paper concludes with some remarks on future directions.

## 2 Staffing Model

Planning staffing levels requires balancing two conflicting goals: minimising cost and maintaining high service levels. Here the service level is defined by a waiting time target. Increasing staff will reduce patient waiting times but will result in high resource costs. A health care system will exhibit stochastic behaviour such

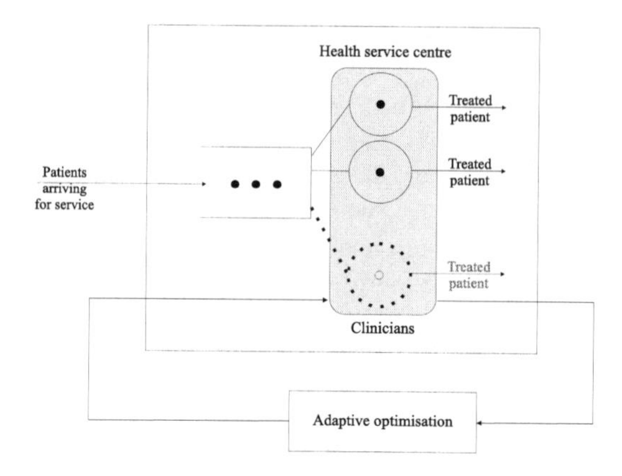


**Fig. 1.** Adaptive Staffing Model

as the uncertainty in how long it takes to treat individual patients and arrival rates of patients to different departments within a hospital. To account for this randomness in behaviour the health system is modelled using queueing theory principles (Fig. 1). The system is modelled as a network of $c(n)$ service stations or number of clinicians at $n$-th time instant. Patients enter the system and either wait in a queue or go for treatment to one of the service stations. This paper considers a First-Come-First-Served (FCFS) discipline and no priority levels between customers even though more complex definitions are possible (and should be incorporated to reflect phenomena found in real systems). The staff planning involves determining how many clinicians are needed to provide a target level of service at minimum cost. The stochastic nature of arrivals and service times are represented by probability distributions. It is assumed that the arrivals are random and Poisson distributed and the service rate is denoted as $\mu(n) = \mu \cdot c(n)$. $\mu$ is the rate of service and $c(n)$ is the number of servers and the staff allocation amounts to deciding the level $c(n)$. A maximum limit to the number of servers that can be added is set to $K$. A $GI/G/c(n)$ queueing model is used here to provide a good approximation of time-varying patient arrival rates and general service time distribution. The results section will consider patient arrival intensities to follow a non-homogeneous Poisson distribution, that is a Poisson process with a time-varying rate parameter to indicate a sudden rush of patients. We will then show that this approach will adaptively allocate staff levels to deal with these dynamics and ensure that patient waiting times do not increase and miss targets.

### 2.1 Transient Approximations for GI/G/c(n) Model

Based on Pollazek-Khintchine formulas and 'transient Little's Laws' [6] the distribution of the patient delay $W(n)$ in the system when he or she arrives at time $n$ is given by,

$$F_n(x) = G \star [1-\rho(n)] \sum_{i=0}^{\infty} \rho(n)^n H_e^{i\star}(x) \tag{1}$$

where $\star$ represents the convolution operation. The system performance parameters at $n$-th time are calculated as follows,

$$\begin{aligned}
\text{Staff utilisation level: } \rho(n)^{-1} &= 1 + \frac{\mu_e}{\mu_1 p(n) + \mu(Q(n) - B(n))} \\
\text{Patient waiting time in system: } E[W(n)] &= m + \mu_1 p(n) + \mu[Q(n) - B(n)] \\
\text{Prob. the patient has to wait in queue: } p(n) &= \alpha(B(n)) \\
\alpha(B(n) &= \frac{P(c; B(n)) - P(c-1; B(n))}{P(c; B(n)) - (B(n)/c)P(c-1; B(n))} \\
B(n) &= \max[0, \mu c[\lambda(n) - Q'(n)]] \\
Q'(n) &= (Q(n) - Q(n-1))/\delta
\end{aligned}$$

where the additional parameter $B(n)$ is the number of clinicians seeing patients at a given time and $P(c;B) = \sum_{i=0}^{c} e^{-B} B^i / i!$ [7]. The number of patients waiting for treatment is $Q(n)$ and $\delta$ is the duration of the $n$-th time interval. The mean and equilibrium mean of the service time distribution is given by $m$ and $m_e$ respectively. The means $\mu$ and $\mu_1$ are the means of service time and residual service time distributions when $Q(n) \geq c$ at the time of a new arrival,

$$\begin{aligned}
H(x) &= 1 - \left[\bar{G}_e(x)\right]^{c-1} \bar{G}(x) \\
H_1(x) &= 1 - \left[\bar{G}_e(x)\right]^{c}
\end{aligned} \tag{2}$$

and $H_e(x)$ is the equilibrium distribution associated with $H(x)$ and its mean is denoted $\mu_e$.

## 3 Adaptive Staff Planning Approach

Lets formalise the staff allocation problem now. Under a specific policy $\pi$ a sequence of decisions are defined which include at a specific time $t$ a decision on the number of staff $1 \leq c(t) \leq K$ to be allocated. In order to achieve both a target waiting time and minimise the accumulated cost of staff resources, penalties on missing targets and cost of patient queueing there is a need to add/remove part of the staff depending on the workload and queue length. This is investigated here in terms of a queueing model with many service stations. Given that $W_{\max}$ is the target maximum waiting time and $C_r$ - the individual staff cost per unit time, $C_p$ - the penalty cost per unit time, $C_s$ - cost associated with the queue length per unit time, the objective function is given as,

$$V_\pi = E_\pi \left[ \int_{t=0}^{\infty} (C_r c(t) + C_p \max(E[W(t)] - W_{\max}, 0) + C_s Q(t)) \; dt \right] \tag{3}$$

The above optimisation problem can be solved as a finite horizon dynamic programming problem. The goal of this paper is to determine self-adaptive policies in order to minimise the accumulated cost of operation. The cost criterion is minimised based on the mean waiting time, queue length and number of staff. In this framework staff allocation is altered at the end of discrete-time equidistant time intervals $\delta = 1/\gamma$ with $\gamma = \max(\lambda) + K\mu$ where $\lambda$ is the patient arrival rate. The cost at epoch $n$ is given by,

$$C(i, c(n)) = (C_r c(n) + C_p \max(E[W(n)] - W_{\max}, 0) + C_s i)/\gamma \quad (4)$$

when $Q(n) = i$. The goal of this problem is to recursively find the decision vector $\mathbf{c} = (c(1), \ldots, c(N))$ which finds the minimum cost path using dynamic programming as follows,

$$V(Q(n) = i) = \max_{\mathbf{c}(n) \in (1,\ldots,K)} \left[ C(i, c(n)) + \sum_{\forall j} p_{ij}^{c(n)} V(Q(n+1) = j) \right] \quad (5)$$

The complexity of this type of solution can be very high therefore we use a technique called neuro-dynamic programming [8] to derive a near-optimal solution. Specifically we consider a one-step look ahead approach to approximate the cost-to-go function such that,

$$V_{n+1}(Q(n+1), c) = \sum_{\forall j} p_{ij}^{c(n)} (C_r c + C_p \max(E[W(n+1)] - W_{\max}, 0) + C_s j)/\gamma \quad (6)$$

for $c \in \{1, \ldots, K\}$. The sojourn time for an arrival at the next epoch $E[W(n+1)]$ is calculated using matrix analytic methods when the number of servers is equal to $c$. Then the resource allocation decision for epoch $(n+1)$ is given by,

$$c(n+1) = \min_{c \in \{1,\ldots,K\}} \{C[Q(n), c(n)] + V_{n+1}(Q(n+1), c)\} \quad (7)$$

## 4 Staff Planning in an A&E Unit

A&E departments are being placed under increasing pressure to process a growing number of patients safely and quickly. This is evidenced by the national government target whereby 98% of patients must spend 4 hours or less from arrival to admission, transfer or discharge, as well as an increase in the number of attendances to A&E departments and walk-in centres in England. Concurrently, in 2003 – 6, seven hospital trusts reported one or more A&E departments closed or downgraded, with one new A&E department opening. For the remaining open A&E departments it is important to be able to predict the changes in patient arrivals and take optimal decisions that meet cost and service level targets. There are many studies describing simulations of A&E departments in which either a Poisson arrivals process is assumed or historical attendance is replicated. There has been some success in application of queueing models to A&E departments in particular the studies by [9,10,11] which will have limited success in complex

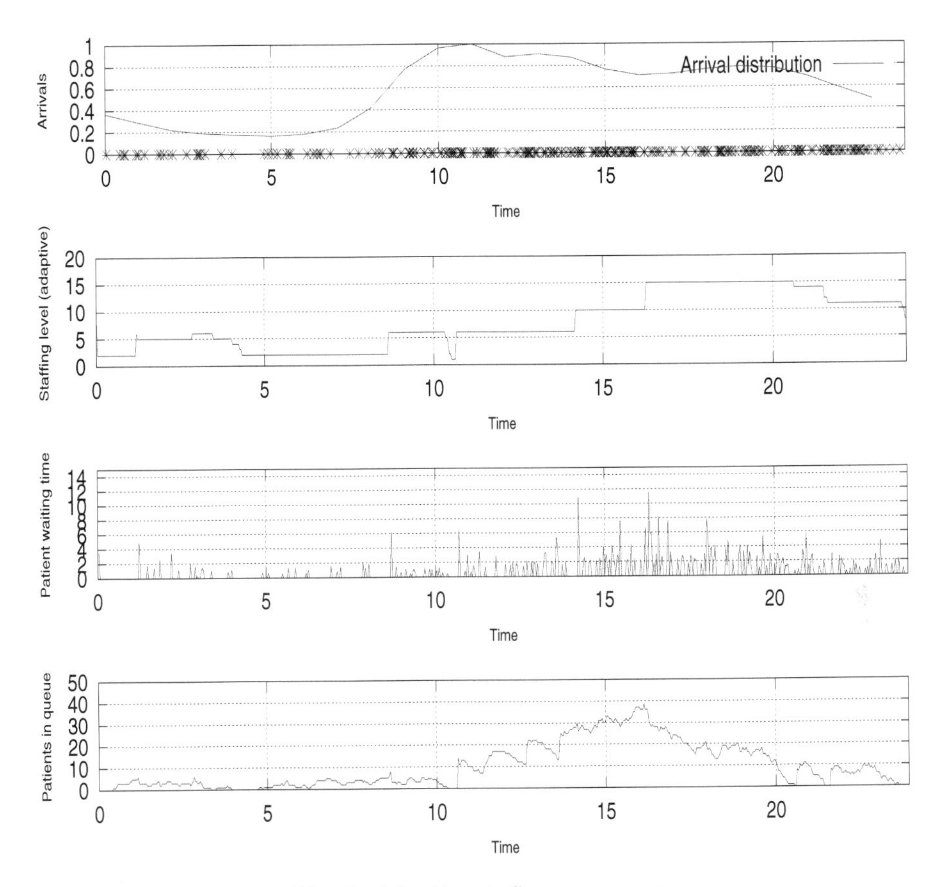


**Fig. 2.** Adaptive staff management

systems. Firstly, the high-level queueing models used typically do not tie system performance to the underlying resources. Secondly, many of these models do not take into account phenomena that occur in the corresponding real life systems such as time-varying arrivals. This paper goes beyond these models and presents two new aspects for the first time. Firstly, the problem formulation jointly finds the optimal set of resources that achieve waiting time targets and minimise cost. Secondly, a transient queueing model and an adaptive performance optimisation technique is developed that allocates resources on-line. Simulation results are presented that shows the adaptive staff planning capabilities of the technique described. A discrete event simulation (DES) of a multi-server queueing system (Fig. 1) is built to model the complete emergency service centre. Here each server represents one staff member. Actual arrivals data from an A&E unit in London obtained over a one year period is used as input to the DES. The arrival rates are computed by averaging over all the days in the year and the arrival rate distribution is given in the first sub-plot in Fig. 2. The service rate of each staff member is assumed to be 1.379 (patients/hour) and the cost parameters are

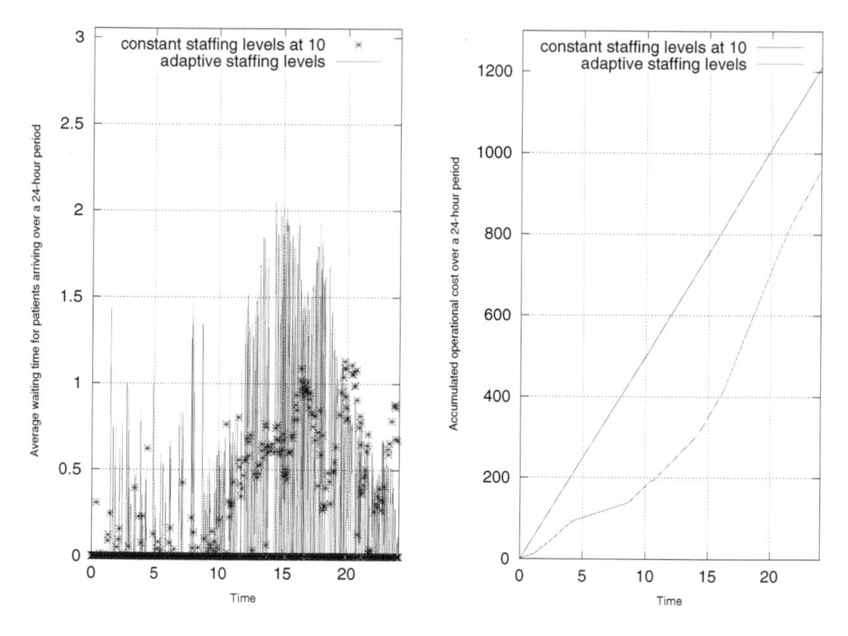


**Fig. 3.** Accumulated cost

assumed to be, (1) cost of patients queueing for treatment is 3 cost units per hour (2) penalty of missing the 4 hour waiting time target is 2 cost units per hour and (3) resource cost is 5 cost units per hour. More realistic parameters can be used and more complex models can be built when these techniques are applied to an actual system. Results are obtained under two conditions over a 24 hour period from $0000-2400$ hours. That is once when the staffing level is changed dynamically and the other when the staffing level is maintained at a constant level of 10 clinicians during the day. The adaptive staffing policy takes values from the set $\{1, 2, 3, 4, 5, 6, 10, 15, 20\}$ such that accumulated cost of operation over the day is minimised. The results of the adaptive staff allocation techniques is given in Fig. 2 where measures from the system are plotted at time slots of 2 minutes. Additionally it is assumed here that when a staff reduction is recommended by the optimisation technique the DES takes the maximum between the suggested level and currently busy number of staff. This is to represent that treatment in progress is not interrupted. Sub-plots $2-3$ shows that when ever the overall waiting time during a time slot increases sharply the staff allocation policy recommends an increase of staff leading to a subsequent drop in the overall waiting time over the next time slot. Sub-plot 4 gives the number of patients waiting in a queue at each time slot during the day. The targeted performance results of using this techniques is given in Fig. 3 and shows the performance management capabilities that can be obtained by adaptively controlling the staffing level. The first sub-plot indicates the average waiting time a patient arriving at the emergency unit would face at any given time during the day. The second sub-plot gives the overall accumulated cost of operating the facility for the day based on the cost parameters given earlier. This highlights the benefits of reduced overall cost when adapting the staffing level dependent upon the current workload and waiting times. This in effect also reduces the under utilisation of expensive staff resources when the emergency unit is not busy.

## 5 Conclusions

This paper presented a new technique to adaptively allocate staff in a IHC system in order to meet quality of service targets and minimise cost. The results obtained show the benefit to management and administration of a health centre to evaluate the effects of staffing policies on performance and cost. Furthermore the unique adaptive nature of this technique means that decisions can be taken in real-time in response to changes in workload.

## References


1. Darzi, A.W.: Ideas from Darzi:polyclinics. NHS Confederation Publications (2008)
2. Bazzoli, B.J., Chan, B., Shortell, S., D' Aunno, T.: The financial performance of hospitals belonging to health networks and systems. Inquiry 37(3), 234–252 (2000)
3. Smith-Daniels, V.L., Schweikhart, S.B., Smith-Daniels, D.E.: Capacity management in health care services: Review and future research directions. Decision Sciences 19, 889–918 (1988)
4. Brailsford, S.C., Lattimer, V.A., Tamaras, P., Turnbull, J.C.: Emergency and on demand health care: modelling a large and complex system. Journal of the Operational Research Society 55, 34–42 (2004)
5. Gorunescu, F., McClean, S.I., Millard, P.H.: A queueing model for bed occupancy management and planning of hospital. Journal of the Operational Research Society 53, 19–24 (2002)
6. Riaño, G.: Transient behaviour of stochastic networks: Application to production planning with load dependent lead times, Ph.D. Thesis, Georgia Institute of Technology (2002)
7. Grassmann, W.K.: Finding the right number of servers in real-world queueing systems. Interfaces 2, 94–104 (1988)
8. Bertsekas, D.P., Tsitsiklis, J.: Neuro-dynamic programming. Athena scientific, Belmont (1996)
9. Coats, T.J., Michalis, S.: Mathematical modelling of patient flow through an Accident and Emergency department. Emergency Medicine Journal 18, 190–192 (2001)
10. Mayhew, L., Carney-Jones, E.: Evaluating a new approach for improving care in an Accident and Emergency department: The NU-care project. Technical report, Cass Business School, City University (2003)
11. Mayhew, L., Smith, D.: Using queueing theory to analyse completion times in Accident and Emergency times in the light of the government 4-hour target. Technical report, Cass Business School, City University, Actuarial Research Paper No. 177 (2006)

# "Do Users Do What They Think They Do?"– A Comparative Study of User Perceived and Actual Information Searching Behaviour in the National Electronic Library of Infection

Anjana Roy[1,2], Patty Kostkova[1], Mike Catchpole[2], and Ewart Carson[1]

[1] City ehealth Research Centre and Centre for Health Informatics, City University, Northampton Square, London UK
[2] Health Protection Agency, 61 Colindale Avenue, London, NW9 5EQ
Anjana.Roy@hpa.org.uk, patty@soi.city.ac.uk, Mike.Catchpole@hpa.org.uk, e.r.carson@soi.city.ac.uk

**Abstract.** In the last decade, the Internet has profoundly changed the delivery of healthcare. Medical websites for professionals and patients are playing an increasingly important role in providing the latest evidence-based knowledge for professionals, facilitating virtual patient support groups, and providing an invaluable information source for patients. Information seeking is the key user activity on the Internet. However, the discrepancy between what information is available and what the user is able to find has a profound effect on user satisfaction. The UK National electronic Library of Infection (NeLI, www.neli.org.uk) and its subsidiary projects provide a single-access portal for quality-appraised evidence in infectious diseases. We use this national portal, as test-bed for investigating our research questions. In this paper, we investigate actual and perceived user navigation behaviour that reveals important information about user perceptions and actions, in searching for information. Our results show: (i) all users were able to access information they were seeking; (ii) broadly, there is an agreement between "reported" behaviour (from questionnaires) and "observed" behaviour (from web logs), although some important differences were identified; (iii) both browsing and searching were equally used to answer specific questions and (iv) the preferred route for browsing for data on the NeLI website was to enter via the "Top Ten Topics" menu option. These findings provide important insights into how to improve user experience and satisfaction with health information websites.

**Keywords:** Digital Library, User Perceived and Actual Behaviour, Evaluation.

## 1 Introduction

Over the last decade, the Internet has become a ubiquitous medium essential for many activities in our daily lives. The healthcare sector has been experiencing a radical transformation in recent years as a result of this unprecedented growth of information technology. The internet has created new opportunities for better and more timely

P. Kostkova (Ed.): eHealth 2009, LNICST 27, pp. 96–103, 2010.

delivery of healthcare, but also brought challenging issues and different responsibilities for medical information providers [1].

Recent researches have investigated whether users have been able to locate information on a web site and are satisfied with content quality rather than investigating user web information seeking behaviour and satisfaction with the ease of location of information. This latter aspect of information delivery is particularly important in the case of healthcare web sites, when information is often being sought on topics that are beyond the everyday experience of the seeker.

Combining qualitative and quantitative methods provides an opportunity to understand the relationship between user perceived and actual navigation behaviour and how that impacts on the user success and satisfaction in information retrieval. We have used this approach with respect to the users of the National electronic Library of Infection (www.neli.org.uk) – a UK national portal for professionals in infection and public health that we are responsible for developing. The results reveal fundamental misunderstanding of web navigation terminology and unexpected differences between user perceived satisfaction and success / failure of navigation.

### 1.1 'User Perceived' versus User Actual Information 'Searching Behaviour'

While there are a number of definitions of user information seeking and searching behaviour, for our purposes we shall use the term *information searching behaviour* as defined by T. Wilson, as all user activity on the web site with the purpose of finding certain information, as opposed to "surfing" the web site without a prior information need [2]. Essential indicators of the probability of knowledge discovery and overall user satisfaction and site usage are determined by (i) whether users use the site the way the site designers expect; (ii) whether they understand terminology used for site searching and (iii) how they navigate the site to find what they are looking for. This is of particular concern in the healthcare domain since failure to locate relevant information, or worse the location and use of out of date or poor quality information can have serious health consequences. Navigation can be measured by so called "disorientation" [3] – loss of a sense of orientation in relation to the web site space. Disorientation can be caused by complexity of the site navigation (browsing and searching access points), unclear terminology, and poor knowledge of the domain [4]. We measured user disorientation by their perceived satisfaction with the: (i) success in locating the most recent information; (ii) ease with which they located the information they were seeking; and (iii) general understanding of the site navigation terminology. The research questions addressed were:

1. Was there any difference between perceived (reported) user information searching behaviour and the actual (observed) behaviour?
2. To find the information, did users seek the information by browsing, searching or combined methods?
3. Did users have insight into the navigation methods they use to locate information
4. Did users find the clinical information they were looking for?
5. Is there any information searching behaviour pattern typical of certain types of users.

## 2 NeLI Navigation Structure

NeLI (http://www.neli.org.uk ), is one of the Specialist Libraries of the National Library of Health (NLH), providing the best available evidence for clinicians, communicable disease control nurses, environmental health officers, public health specialists and others around prevention, treatment, investigation and management of infectious diseases. The information is available through the NeLI portal using two major navigation methods for Information Searching.

**Browsing** includes the following: (i) comprehensive A to Z listing of diseases/organism – based on MESH keywords including diseases, organisms and symptoms; or a filtered browse restricted to (ii) Top 10 Topics based on commonly accessed topics of the website 2001 - 2004; (iii) Factsheets ; (iv) Guidelines; (v) Antimicrobials & Antimicrobial Resistance or (vi) List of Infectious Disease Society Websites

**Searching:** Drop-down key word search: MESH keyword-based menu. The drop-down search menu is based on the most-frequently used MESH keywords around infection and public health including Diseases, Organism, and Symptoms. As MESH does not include a taxonomy for management of disease from a public health perspective, it has been extended to cover the appropriate terms. All NeLI pages are mapped to these terms to enable precise knowledge retrieval.

**Free-text search:** In addition, free text search is supported to allow searching for any keywords – in this case, full text search, powered by Free Find, is used to retrieve appropriate pages.

## 3 Information Searching Investigation Methods

We developed an evaluation methodology of the NeLI users, to investigate (i) their information searching behaviour, (ii) perceived and actual behaviour; (iii) satisfaction with the information on the site; and (iv) understanding users insight into the navigation methods used by them.

**Study Design**
An on-line questionnaire survey and weblog analyses were combined to determine actual and perceived user behaviour. IP addresses of respondents were used to track their navigational pathway. The pathway answers from the questionnaire were compared with their weblogs.

Users were asked to have a clinical or non-clinical question in mind before navigating the site. This was essential to evaluate user information searching as opposed to surfing. The target user group were NeLI professionals who regularly visit the site. The majority of the users were confident Internet users.

To overcome the limitations of questionnaire surveys and web logs when used alone, we combine these two methods to gain a deeper understanding of user information searching behaviour, satisfaction and perception. The results of the online questionnaire addressing the navigation, perception and user needs were compared with user actual behaviour from the web logs.

**Online Questionnaire:** Questionnaires offer an standardised means of collecting information about people's knowledge, beliefs, and attitudes. Questionnaires were used to test user knowledge/attitude change before and after use of the website [5], while exit questionnaires were used to investigate user satisfaction [6]. This study determines whether there was a change in knowledge after visiting the site.

**Web logs:** Web logs represent a recording of user behaviour while visiting a website, and provide invaluable insights into the users' actual behaviour and the time, date and type of information the user has accessed. The drawback of web logs is that while they can reveal essential information about user navigation, they cannot provide insight into why users behaved the way they did and how satisfied they were with their information searching results.

**Combining the methods:** We looked at the declared navigational pathway preference, searching , browsing, combined method etc - (from questionnaires) and the actual path followed by the users (from web logs) for 15 users.

Details of searching and browsing behaviour were analysed for evidence of correlations between types of 'user perceived' or 'reported' and 'actual preferences' or 'observed' behaviour. We also assessed the web logs for evidence of searching or browsing for information other than that declared as the primary subject of interest on the questionnaire.

# 4 Study Results

The data collection part of this study took six weeks (August – September 2004). Nineteen users took part in the study, but we were able to identify web log data for only 15 of the users.

In the questionnaire, participants were specifically asked to record the question they were attempting to answer. For each user we compared their opinion on whether (a) they had found the answer to their question and (b) how in their view they navigated the site to get the answers and compared that to the web log evidence on (i) whether they found the most appropriate information and (b) how they actually performed the information search with regard to their question. In a majority of cases the respondents answered the their specific question in the same session as while they looking for their answers, this did not provide the opportunity to determine whether questionnaire recall accuracy was affected.

## 4.1 Did the Users Find the Information They Were Looking for?

Our results indicated that all users "reported" to have found the answers to the question they specifically were attempting to address. Our 'observed' result indicate that all except one of our users did access the page of interest specified. The user whose "reported" and "observed" page did not coincide did visit "Other websites of interest" in NeLI. So it is possible that the user did locate their topic of interest via the NeLI website – but our weblog data are unable to confirm this.

## 4.2 Navigation on NeLI: Browsing and Searching

Users were asked whether they preferred browsing, searching or combined methods for accessing information on NeLI and to provide details of the types of browsing and searching employed.

**Table 1.** Users' preference about the mode of navigation as answered in question one of the questionnaire. The questionnaire and web log answers were compared by calculating the kappa measure of agreement (Table 1). The measure of agreement was 0.59, meaning moderate agreement between the answers provided in the questionnaire and actual behaviour recorded on the web-logs, with respondents more likely to report, or recall, searching activity than browsing activity.

| Weblog | Questionnaire | | |
|---|---|---|---|
| | Browse (%) | Search(%) | Combination(%) |
| Browse | 20 | 7 | 13 |
| Search | 0 | 40 | 0 |
| Combination | 0 | 7 | 13 |

On the basis of observed (web lob) behaviour, both 'Browse' and 'Search' appear to be equally preferred for navigation to find the answer to the specific question addressed by the user.

## 4.3 Browsing and Searching Behaviour Details

Questionnaire responses on "Browse behaviour" indicated that more users chose to browse a 'restricted' view of content, selecting Fact sheets, Guidelines and Top Ten topics, than use the full A-Z listing and List of Infectious Diseases. However, this preference was not statistically significant. Overall, most of the users reported their method of navigation accurately, when compared to web logs. For "searching" the majority of users reported behaviour also coincided with their 'observed' behaviour (Table 2). A wide range of options were used in the 'pull down options' in search. These varied from searching information on specific viruses e.g. ***Cytomegalovirus***, seeking information on travel associated illnesses to looking information for contributors. None of the users in this study used the "Free text search".

The numbers in each column do not add up to 100, because users indicated that they used a combination of options in Browse along with the search options.

These results indicate the following:

1. Users were successful, and perceived themselves to be successful, in locating the information they were seeking.
2. Most users located information using a combination of navigation options; in the majority of the cases this has been reported accurately.
3. There is no significant preference for any particular browse option (other than that 'Antimicrobial Resistance' was not used at all in this study).
4. The "Pull down" menu in the search option is the only search option used in this study – "Free-text" was not used at all.

**Table 2.** Comparison of reported and observed browse and search behaviours

| | | Questionnaire (percent) | Web logs (percent) |
|---|---|---|---|
| Browse | | | |
| | A to Z listing of pages on NeLI | 7 | 13 |
| | Top 10 Topics | 13 | 20 |
| | Factsheets listed on NeLI | 13 | 13 |
| | Guidelines listed on NeLI | 13 | 13 |
| | Antimicrobials Resistance | 0 | 0 |
| | List of Infectious Disease Society Websites | 7 | 7 |
| Search | | | |
| | Pull down Menu | 67 | 60 |
| | Free Text Search | 0 | 0 |

## 4.4 Cases Where the Users Reported Navigation Behaviour and Observed Behaviour Did Not Match

When the 'observed' and 'reported' results did not match – it was due to one of the following reasons:

1. Users were 'observed' to use a combination of navigational options while reporting only one option to access the page of interest.
2. Users reported using a different navigation technique to that observed e.g they reported to be using 'search' – but were observed to reach the page of interest via 'Browse' options.

80% of the users reported that they found navigation of the site easy while 20% reported it to be difficult. The 20% expressing difficulty 'reported' their method of navigation correctly, i.e. their 'observed' and 'reported' behaviour matched perfectly.4.5. Navigation to "Other Pages of Interest"

The web log data were able to inform us of the information searching method used when users visited "other pages of interest" in this session - typically accessed after answering their "specific question". Web log analysis indicates that the most favoured method of navigation in this case was Top 10 Topics, which also matches our specific question web log data (Table 3). Of the users, majority (93%) used the browsing option to visit other pages of interest, few (7%) used the search 'Drop Down – key word search', with no-one using the Free Text Search. We have not asked users for their navigation methods for seeking information other than that used to answer their clinical question, as we did not expect such a high percentage of users to "surf" while answering the questionnaire. Therefore, we do not have the questionnaire data for comparison in this case, but did find the web log data very revealing. We can conclude that for 'surfing ' through the site Top 10 Topics is the favoured option.

Please note : The total percent of users does not add up to 100 as this table shows the total number of people who actually went to "other" pages of interest – while 19% only answered the questionnaires during the session.

**Table 3.** Navigational method used to access the "other pages of interest"

| | | **Web logs** (percent) |
|---|---|---|
| **Browse** | | |
| | A to Z listing of pages on NeLI | 7 |
| | Top 10 Topics | 27 |
| | Factsheets listed on NeLI | 13 |
| | Guidelines listed on NeLI | 7 |
| | Antimicrobials & Antimicrobial Resistance | 7 |
| | List of Infectious Disease Society Websites | 13 |
| **Search** | | |
| | Drop Down - Key word search | 7 |
| | Free Text Search | |

## 5 Discussion

Development of successful healthcare Web sites demands regular evaluation to provide better understanding of the underlying issues of user satisfaction with the provided information and ease with which it can be located. In this paper we investigated the discrepancy between user perceived and actual navigation behaviour.

In this study we have correlated the pattern of navigation of individual users with their views on the NeLI web site and analysed the discrepancies between the sets of data. Choo et al. have used questionnaires and web logs and developed a modified model to describe common repertoires of information seeking by users in general [7]. Here, we have tried to determine whether the users have reported their navigation strategy correctly – or have become disorientated, which can arise from unfamiliarity with the structure or conceptual organisation of the site . The decision regarding which web page to view next involves understanding one's current location within the site, then selecting the proper route .

The users were able to locate the information they were seeking specifically from the NeLI website. This implies that largely the users understand the terminology used for navigating the site, and that they use the site as the developers expect them to.

However, the 'user perceived' and the actual user behaviour did not always tally. There appears to be slight confusion regarding the search and browse options, with a few users using a combination of options to access the information they were seeking, but failing to realise this, i.e. they reported only using one navigation methodology. Web-space disorientation could be a likely explanation. Therefore, efforts are being made to make the browse and search access points clearer and more simple to use in the next version of the NeLI.

In this study we wanted to determine the preferred method of navigation of the NeLI site. This appeared to depend on :

1. whether the user was trying to answer a specific clinical/non-clinical question using the NeLI website. We obtained this information primarily from the web logs study, which indicated a preference for searching to locate information. When searching, only the "pull down" menu was used – free text was not used at all. Where browse was used, there was little evidence of a preferred point of entry, although users did seem to be more likely to browse within a particular type

of document ('Factsheet', 'Guidelines', etc) than opt for a subject-based filter as the starting point for browsing.

2. When the users were just surfing though NeLI their preference was to browse, using "Top 10 Topics" as a starting point i.e. they appeared to be content to be guided by the website authors, or other users, as to where they might find the most useful information. In this case we only had information on "observed" method of navigation. The popularity of this method of navigation has led the developers to research this area further and broaden the scope of this option. Currently the NeLI developers are in the process of identifying the Top Twenty Five Topics in infectious diseases.

The users found the site to be too "wordy" and suggested improvement by addition of clips, videos and images. This information is invaluable for the site developers and obviously cannot be obtained from any web log analysis. This study will enable web developers to improve navigation system within their systems.

## 6 Conclusion

By combining qualitative and quantitative methods (online questionnaires and weblogs) we could demonstrate that there is reasonable correlation between user perceived information seeking behaviour and the actual behaviour. This pilot study was part of a long-term project investigating impact evaluation of healthcare Web sites, and the development of validated tools for providing feedback that can inform website improvements.

## References

1. Armstrong, R.: Appropriate and effective use of the Internet and databases. Clin. Rheumatol. 22(3), 173–176 (2003)
2. Wilson, T.D.: Human Information Behaviour. Information Science 3(2) (2000); special Issue on Information Science Research
3. Juvina, I.: The Impact of Link Suggestions on User Navigation and User Perception. In: Ardissono, L., Brna, P., Mitrović, A. (eds.) UM 2005. LNCS (LNAI), vol. 3538, pp. 483–492. Springer, Heidelberg (2005)
4. Draper, S.W.: Supporting use, learning, and education. Journal of computer documentation 23(2), 19–24 (1999)
5. Madle, G., Kostkova, P., Mani-Saada, J., Williams, P., Weinberg, J.R.: Changing public attitudes to antimicrobial prescribing: can the Internet help? Informatics in Primary Care 12(1), 19–26 (2004)
6. Huntington, P., Williams, P., Nicholas, D.: Age and gender user differences of a touchscreen kiosk: a case study of kiosk transaction log files. Informatics in Primary Care 10(1), 3–9 (2002)
7. Choo, C.W., Detlor, B., Turnbull, D.: 2000ormation seeking on the web– An integrated Model of browsing and searching. First Monday 5(2) (February 2000), http://firstmonday.org/issues/issue5_2choo/index.html

# MEDEMAS -Medical Device Management and Maintenance System Architecture

Ülkü Balcı Doğan[1], Mehmet Uğur Doğan[1,2], Yekta Ülgen[1], and Mehmed Özkan[1]

[1] Boğaziçi University, Biomedical Engineering Institute, İstanbul, Turkey
[2] TÜBİTAK, National Research Institute of Electronics and Cryptology, Turkey
{ulkub,mugurd,ulgeny,mehmed}@boun.edu.tr

**Abstract.** In the proposed study, a medical device maintenance management system (MEDEMAS) is designed and implemented which provides a data pool of medical devices, the maintenance protocols and other required information for these devices. The system also contains complete repair and maintenance history of a specific device. MEDEMAS creates optimal maintenance schedule for devices and enables the service technician to carry out and report maintenance/repair processes via remote access. Thus predicted future failures are possible to prevent or minimize. Maintenance and repair is essential for patient safety and proper functioning of the medical devices, as it prevents performance decrease of the devices, deterioration of the equipment, and detrimental effects on the health of a patient, the user or other interacting people. The study aims to make the maintenance process more accurate, more efficient, faster and easier to manage and organize; and much less confusing. The accumulated history of medical devices and maintenance personnel helps efficient facility planning.

**Keywords:** medical device, medical device maintenance, maintenance protocol, scheduling.

## 1 Introduction

World Health Organization defines a medical device as "any instrument, apparatus, implement, machine, appliance, implant, in vitro reagent or calibrator, software, material or other similar or related article, intended by the manufacturer to be used, alone or in combination, for human beings for one or more of the specific purposes of; diagnosis, prevention, monitoring, treatment or alleviation of disease or injury; investigation, replacement, modification, or support of the anatomy or of a physiological process; supporting or sustaining life [1]".

It is mentioned in the same document that safety and performance of medical devices should be continually assessed when they are in use, since these characteristics can only be proven if one measures how a device stands up in these conditions. It is not possible to predict all possible failures or incidents caused by device misuse with pre-marketing review processes. It is through actual use

P. Kostkova (Ed.): eHealth 2009, LNICST 27, pp. 104–107, 2010.

that unforeseen problems related to safety and performance can occur[1]. Joint Commission International Accreditation sets the standard for medical equipment and utility systems as: "The organization plans and implements a program for inspecting and maintaining medical equipment and documenting results[2]".

The aim of this research is to design and develop medical device maintenance management software which will keep record of medical devices, their information, maintenance protocols and repair/maintenance histories; MEDEMAS assigns the foregoing maintenance dates and the technicians responsible for them; informs the relevant technicians and the users of the medical device. The required tools and devices to carry put the task are also scheduled. The architecture makes it possible to carry out the maintenance process remotely.

It has been reported in the literature that successful applications of various maintenance optimization models are rare. Computational difficulties; difficulties of collection of data and modeling of failure distribution; and the gap between theory and practice are the major problems in applying these models. To close the gap between theory and practice, an in-depth investigation is essential to develop an effective methodology for modeling equipment's failure distribution or degradation process. Some key points are what information is required for modeling, how such information is obtained, how the model parameters are estimated, and how the model is updated when new information becomes available [3].

## 2 Design Considerations

The most comprehensive and challenging side of the study is the optimization of maintenance scheduling. We intend to make the study a well constructed preventive maintenance. All actions carried out on a planned, periodic, and specific schedule to keep an item/equipment in stated working condition through the process of checking and reconditioning [4].

While planning maintenance scheduling, the number and the availability status of maintenance equipment should be taken into consideration. Similarly, the number, status, and expertise of technicians; availability of medical device to be maintained and its maintenance history are also required. In case of failure the action to be taken is to adjust a new maintenance schedule after approximating repair period; meanwhile the technician will be directed to a repair request form for the medical device. These operations should not be done manually; underlying software should make maintenance scheduling and the related jobs automatically.

The application has a powerful evaluating facility. Evaluation of the medical devices, the technicians and the maintenance processes can be calculated automatically using database scoring. These databases contain grading of processes/properties which will be added up to make evaluations. Keeping information of and evaluating the technical staff is especially important as a significantly large proportion of total human errors occur during the maintenance phase; however, human error in maintenance has not been given the attention it deserves [5].

## 3 System Architecture

In MEDEMAS, a main server is remotely accessed by technicians with PDAs through GPRS/WLAN. Main server can be accessed by hospital personnel for data entry, management and reporting facilities. Computers through either internet or intranet can be used. The management system topology can be seen in Fig. 1.

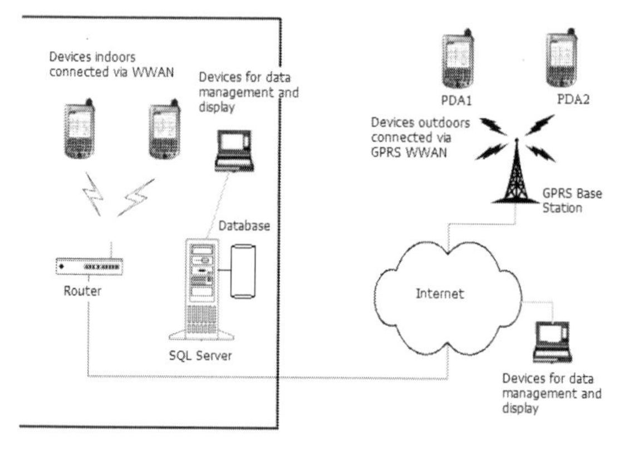


**Fig. 1.** Typical remote access management scenario

There are two main types of access to the application. First one is remote maintenance access which is used by the technician on process. The access is limited with technician's authorities, and his intervention (update/delete records etc.) is limited. Main process is filling out the maintenance form and sending it to the main server. Besides this, the technician is supplied with the maintenance protocol and the maintenance history of the device; he is informed about the jobs ahead he is responsible for and is alerted for incomplete jobs.

Direct application management is the second access type. Both internet and intranet can be used, and almost all data manipulation, reporting, display, evaluation and tracking are via this type. It can be further divided into three: The

**Table 1.** Anaesthetic ventilator Standard and Measurement Values, a section from maintenance form/database. Values and criteria are derived from the reference given.

| Criteria[6] | Allowed/Recommended Values[6] | Measured Values |
|---|---|---|
| Maximum mains voltage | 250 V | |
| Maximum case leakage current | 100 $\mu$A | |
| Maximum patient leakage current | $< 320\mu$A | |
| Maximum earth leakage current | 5 mA | |
| Operating Temperature | 5°C to 45°C | |

first process is forming databases of all devices, maintenance protocols, technicians, trainings, maintenance equipment, maintenance calendar, etc. Recording, update and deletion are main processes in terms of data management. The measurements and controls will be made for compliance with International/European Medical Device Standards. Some standard values and test values are given as sample in Table1 above.

The second process is displaying the database; it includes construction of many reports. Preparing proper reports, it will be possible to track the devices in the hospital, the status of devices, the most problematic devices, the problems experienced, the cost of individual devices in terms of maintenance, etc. The reporting process helps the directors in following the medical equipment, detecting the devices most problematic or most expensive to maintain, finding out the most experienced problems and the reasons of problems. Similarly, it is possible to track the technical staff. With proper reports, it is possible to track technicians, their performances, their success rates and timings; evaluation of technicians can be made.

The third process is designing and implementing an optimal maintenance schedule. An underlying program runs in cases of maintenance entry or update. A change in technician record or maintenance accessory record also causes the program to run. The program creates a new schedule, optimizing parameters like technician, maintenance device, medical device availability, recommended device maintenance period, approximate repair duration, historical data about maintenance activity of the medical device, etc.

## 4 Conclusion Remarks and Future Work

The application architecture of MEDAMAS is carefully built. Database creation and data management modules are completed. Databases are supplied with the flexibility to allow automatic processes. Underlying automation and remote-connection applications are still under construction.

## References

1. World Health Organization, Medical Device Regulations: Global overview and guiding principles. Geneva (2003)
2. Joint Commission Internal Accreditation of Healthcare Organizations, Hospital Accreditation Program. FMS.7 (2000)
3. Tsang, A.H.C., Yeung, W.K., Jardine, A.K.S., Leung, B.P.K.: Data management for CBM optimization. J. Qua. Maint. Eng. (1), 37–51 (2006)
4. Dhillon, B.S.: Engineering Maintenance: A Model Approach. CRC Press, Boca Raton (2002)
5. Dhillon, B.S., Liu, Y.: Human error in maintenance: a review. J. Qua. Maint. Eng., 21–36 (2006)
6. ISO, Inhalational anaesthesia systems: Anaesthetic ventilators, ISO 8835-5 (2004)

# ROC Based Evaluation and Comparison of Classifiers for IVF Implantation Prediction

Asli Uyar[1,⋆], Ayse Bener[1], H. Nadir Ciray[2], and Mustafa Bahceci[2]

[1] Bogazici University, Department of Computer Engineering, Istanbul, Turkey
{asli.uyar,bener}@boun.edu.tr
[2] German Hospital and Bahceci IVF Center, Istanbul, Turkey
{nadirc,mbahceci}@superonline.com

**Abstract.** Determination of the best performing classification method for a specific application domain is important for the applicability of machine learning systems. We have compared six classifiers for predicting implantation potentials of IVF embryos. We have constructed an embryo based dataset which represents an imbalanced distribution of positive and negative samples as in most of the medical datasets. Since it is shown that accuracy is not an appropriate measure for imbalanced class distributions, ROC analysis have been used for performance evaluation. Our experimental results reveal that Naive Bayes and Radial Basis Function methods produced significantly better performance with (0.739 $\pm$ 0.036) and (0.712 $\pm$ 0.036) area under the curve measures respectively.

## 1 Introduction

In-vitro fertilization (IVF) [1] is a common infertility treatment method during which female germ cells (oocytes) are inseminated by sperm under laboratory conditions. Fertilized oocytes are cultured between 2-6 days in special medical equipments and embryonic growth is observed and recorded by embryologists. Finally, selected embryo(s) are transferred into the woman's womb. Predicting implantation (i.e. attachment of the embryo to the inner layer of the womb) potentials of individual embryos may expedite and enhance expert judgement for two critical issues: 1) the decisions of number of embryos to be transferred and 2) selection of embryos with highest reproductive viabilities. Increasing the number of transference embryos may increase the pregnancy probability but also increase possible complications arising from multiple pregnancies. Therefore, in clinical practice, it is desired to transfer minimum number of embryos with the highest implantation potentials. In this study, implantation prediction is considered as a binary classification problem such that the embryos have been classified into two categories as implants and no-implants.

The recent literature presents applications of machine learning methods in IVF domain. Case-based reasoning system [2], neural networks [3], decision tree models [4] [5] [6] and Bayesian classification system [7] were utilized for prediction

⋆ Corresponding author.

P. Kostkova (Ed.): eHealth 2009, LNICST 27, pp. 108–111, 2010.

of IVF outcome. However, direct comparison of these studies is not possible due to variety of input features, training and testing strategies and performance measures. The specific aim of this paper is to determine the best classification algorithm for implantation prediction of IVF embryos.

## 2 IVF Dataset

We have constructed a dataset from a database including information on cycles performed at the German Hospital in Istanbul from January 2007 through August 2008. Dataset included 2453 records with 89% positive and 11% negative implantation outcomes. Each embryo was represented as a row vector including 18 features related to patient and embryo characteristics (Table 1).

**Table 1.** Selected dataset features for each embryo feature vector

| Dataset Features | Data Type |
|---|---|
| ***Patient Characteristics*** | |
| Woman age | Numerical |
| Primary or secondary infertility | Categorical |
| Infertility factor | Categorical |
| Treatment protocol | Categorical |
| Duration of stimulation | Numerical |
| Follicular stimulating hormone dosage | Numerical |
| Peak Estradiol level | Numerical |
| Endometrium thickness | Numerical |
| Sperm quality | Categorical |
| ***Embryo Related Data*** | |
| Early cleavage morphology | Categorical |
| Early cleavage time | Numerical |
| Transfer day | Categorical |
| Number of cells | Numerical |
| Nucleus characteristics | Categorical |
| Fragmentation | Categorical |
| Blastomeres | Categorical |
| Cytoplasm | Categorical |
| Thickness zona pellucida | Categorical |

## 3 Experiments and Results

Six classification algorithms were applied: the Naive Bayes, k-Nearest Neighbor (kNN), Decision Tree (DT), Support Vector Machines (SVM), Multi Layer Perceptron (MLP) and Radial Basis Function Network (RBF) [8]. Two-thirds of the dataset was randomly selected for training a predictor model and the remaining one-third was utilized for testing. The random train set and test set partitioning was repeated 10 times avoiding sampling bias. The reported results were the mean values of these 10 experiments. A t-test with P = 0.05 was applied in order to determine the statistical significance of results.

Various sampling strategies have been applied in order to overcome the imbalanced data problem such as over-sampling and under-sampling. In a recent study, Maloof showed that Receiver Operating Characteristics (ROC) curves produced by sampling strategies are similar to those produced by varying the decision threshold [9]. ROC analysis provides reliable evaluation of classifier performance considering True Positive Rates (TPR) and False Positive Rates (FPR). All the values of TPR and FPR have been calculated by varying the decision thresholds in the range of [0:0.05:1]. The resulting set of (TPR (sensitivity), FPR (1-specificity)) pairs are plotted as a 2D ROC curve. The upper left point (0,1) on the ROC curve represents perfect classification. Therefore, the threshold value that gives the nearest point to (0,1) is accepted as the optimum decision threshold ($t_{opt}$).

We have used Weka machine learning tool [10] to perform classification. Among six methods, Naive Bayes and RBF were significantly better. The results were plotted as ROC curves, which appear in Figure 1 demonstrating the effect of threshold optimization. Naive Bayes classification with optimized threshold results in 67% sensitivity and 30% false alarm rate.

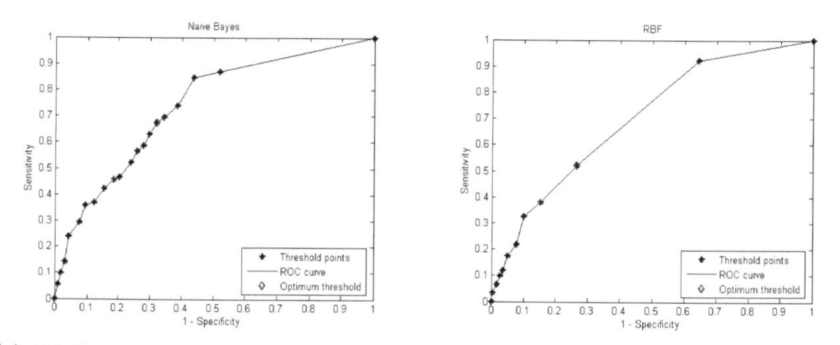


(a) ROC curve for Naive Bayes classification. AUC is 0.739 ± 0.036

(b) ROC curve for RBF classification. AUC is 0.712 ± 0.036

**Fig. 1.** ROC analysis representation of Naive Bayes and RBF classifiers for IVF dataset

## 4 Conclusions and Future Work

In this paper, we have presented a novel study predicting implantation potentials of individual IVF embryos. We have constructed an original dataset and performed predictions using six classification methods. Experimental results show that Naive Bayes and RBF produce significantly better performance for the implantation prediction problem. Model optimization and classifier comparison have been performed using ROC analysis.

To be applicable in clinical practice, the machine learning algorithms have been investigated in a comprehensive and comparative manner allowing reliable selection of best fitting algorithm. In machine learning applications, it is crucial to deal with biases arising from training-testing strategies. We have applied stratified cross validation in order to overcome the sampling bias. Statistical validity is investigated by conducting t-tests. Six models from important representatives of diverse algorithms (statistical classifiers, decision tree approaches, neural networks, support vector machines and nearest neighbor methods) have been applied. Since the data used in this study comes from a single source, it is crucial to consider the external validity of the presented results. The experiments need to be replicated on different IVF datasets, however, public dataset construction and data sharing has been a major research challenge in this domain.

Future work includes improving the information content of dataset by collecting new input features. Combination of classifiers should also be investigated to construct a more powerful decision support system for IVF domain.

## References


1. Steptoe, P.C., Edwards, R.G.: Birth after re-implantation of a human embryo. Lancet 2, 366 (1978)
2. Jurisica, I., Mylopoulos, J., Glasgow, J., Shapiro, H., Casper, R.F.: Case-based reasoning in ivf: Prediction and knowledge mining. Artificial Intelligence in Medicine 12, 1–24 (1998)
3. Kaufmann, S.J., Eastauh, J.L., Snowden, S., Smye, S.W., Sharma, V.: The application of neural networks in predictingthe outcome of in-vitro fertilization. Human Reproduction 12, 1454–1457 (1997)
4. Saith, R., Srinivasan, A., Michie, D., Sargent, I.: Relationships between the developmental potential of human in-vitro fertilization embryos and features describing the embryo, oocyte and follicle. Human Reproduction Update 4(2), 121–134 (1998)
5. Passmore, L., Goodside, J., Hamel, L., Gonzalez, L., Silberstein, T., Trimarchi, J.: Assessing decision tree models for clinical in-vitro fertilization data. Technical report, Dept. of Computer Science and Statistics University of Rhode Island (2003)
6. Trimarchi, J.R., Goodside, J., Passmore, L., Silberstein, T., Hamel, L., Gonzalez, L.: Comparing data mining and logistic regression for predicting ivf outcome. Fertil. Steril (2003)
7. Morales, D.A., Bengoetxea, E., Larranaga, B., Garcia, M., Franco, Y., Fresnada, M., Merino, M.: Bayesian classification for the selection of in vitro human embryos using morphological and clinical data. Computer Methods and Programs in Biomedicine 90, 104–116 (2008)
8. Alpaydin, E.: Introduction to Machine Learning. MIT Press, Cambridge (2004)
9. Maloof, A.M.: Learning when data sets are imbalanced and when costs are unequal and unknown. In: Workshop on Learning from Imbalanced Data Sets (2003)
10. Witten, I.H., Frank, E.: Data Mining: Practical machine learning tools and techniques, 2nd edn. Morgan Kaufmann, San Francisco (2005)

# Evaluation of Knowledge Development in a Healthcare Setting

Scott P. Schaffer

Purdue University, Regenstrief Center for Healthcare Engineering,
Mann Hall 225, 203 Martin Jischke Drive, West Lafayette, Indiana USA 47907
schaffer3999@gmail.com

**Abstract.** Healthcare organizations worldwide have recently increased efforts to improve performance, quality, and knowledge transfer using information and communication technologies. Evaluation of the effectiveness and quality of such efforts is challenging. A macro and micro-level system evaluation conducted with a 14000 member US hospital administrative services organization examined the appropriateness of a blended face-to-face and technology-enabled performance improvement and knowledge development system. Furthermore, a successful team or microsystem in a high performing hospital was studied in-depth. Several types of data methods including interview, observation, and questionnaire were used to address evaluation questions within a knowledge development framework created for the study. Results of this preliminary study focus on how this organization attempted to organize clinical improvement efforts around quality and performance improvement processes supported by networked technologies.

**Keywords:** Knowledge transfer, Microsystems, Knowledge Systems, Evaluation.

## 1 Introduction

The relationship between knowledge, performance, and quality development within organizations is complex. Knowledge management theory recognizes the importance of efficient information management but greater emphasis is on the creation and sharing of knowledge. There are several reasons for this shift. One is the value organizations place on innovativeness requiring freedom to experiment. Dorothy Leonard-Barton [1] suggests learning organizations are confident in their ability to innovate and solve problems precisely because they "know what they know". Sustainable knowledge creation presents the challenge of finding and reusing what one knows when it is needed.

Knowledge management approaches in healthcare are diverse and often related to increased emphasis on quality of care. Berwick [2] argues that experiential learning by healthcare professionals can be facilitated by local tests and evidence collection using plan-do-study-act cycles. This notion has gained some traction and is seen as an alternative to randomized clinical trials especially relative to understanding process innovations.

P. Kostkova (Ed.): eHealth 2009, LNICST 27, pp. 112–115, 2010.

One way to support knowledge created through experiential learning is to support individuals in their collaborative practice. Teams or units of individuals with technical expertise and prior experience with a particular problem and a sufficient amount of resources have great potential to innovate. For example, a study of 43 hospital teams or microsystems identified eight themes associated with high performance. Those themes are: integration of information, measurement, interdependence of care team, supportiveness of the larger system, constancy of purpose, connection to community, investment in improvement, and alignment of roles and training [3].

Socially mediated knowledge development approaches represent a convergence of knowledge, technology, and design practice situated in communities of practice. One study explored how a community of practice facilitated quality initiatives in surgical oncology [4]. The socio-cultural context in which work is performed, and knowledge is applied, is the subject of such communities. This idea supports the notion that organizations learn as they solve problems [5].

Nonaka [6] suggests that knowledge is not another branch of information technology but rather is a kind of human capital simultaneously produced and consumed. This notion of renewable knowledge suggests a convergence of social-cultural and technical systems. The current evaluation study is grounded in the practice of knowledge development within and across hospital systems.

## 2 Methods

This study was conducted in partnership with a 1400 member, US-based hospital administrative services organization. The overall goal of the study was to determine if technology-enabled Clinical Improvement Services (CIS) support critical performance of hospital staff. Specific questions addressed: What is the theoretical and intellectual basis for the CIS design? Do these services support member performance and quality improvement projects? Do they support member knowledge development processes?

Examples of online CIS approaches include the documentation and dissemination of member hospital best practices related to key clinical indicators; forms, documents, and artifacts; and educational interactions such as podcasts, videoconferences, forums, and coaching sessions.

### 2.1 Study Design

Data collection was grounded in the knowledge development framework shown in figure 1. The framework represents a basic systems approach that describes iterative knowledge building cycles. This framework is conceptually similar to the Plan-Do-Check-Act problem solving cycle developed by Shewhart in the 1930's and popularized by Deming and Juran. Solving problems that impact business processes drives activities in the planning phase. Small teams or microsystems are convened to conduct analyses, and generate and test potential solutions to problems or address opportunities for improvement. Artifacts created during this phase are essential to developing capacity for solving future problems. Selected solutions are validated through testing and evaluation. Process and product artifacts are captured, formatted,

**Fig. 1.** Knowledge development system incorporating knowledge creating (planning, building, evaluating) and knowledge sharing (storing and sharing) processes

tagged, and digitized for storing and retrieval. New knowledge is shared across internal and external networks in phone calls, hallway conversations, formal meetings, emails, forums, and conferences.

Research questions were investigated using literature review, questionnaire, direct observation, and interview. A case study conducted with a high performing hospital team was completed to see the framework in action.

# 3 Findings

**Macro-level.** A major finding based on questionnaire results is that current web-supported CIS were not easily understood by hospitals with less sophistication in quality and performance improvement processes. These hospitals had fewer resources to learn about, adopt, and adapt CIS innovations locally. Use of the CIS by hospital performance or quality improvement staff was more event oriented rather than as a primary resource. This finding suggests that the CIS design was more traditional in that it focused on transactions and artifact storing rather than promoting dialogue and a practice community. Questionnaires revealed that high performing hospitals frequently share successes and practices in informal social and professional networks with in-hospital peers and those in other hospitals. The findings suggest that online knowledge development systems should focus on patient care and related problem solving rather than on knowledge development.

**Micro-level.** Case study findings were interpreted within the framework in figure 1. Study of a high performing hospital in the southeastern United States indicated that a culture of improvement created conditions for strategic resource allocation to support team problem solving. Interviewees used the language of quality and performance improvement frequently, knowledgably, and passionately. This level of engagement was not based on specific clinical improvement services, processes, or tools but rather on patient outcomes. Examples of team problem solving within the knowledge development systems framework include:

Planning: "*...increased awareness of SCIP really drove home the importance of follow through on orders... I had heard of SCIP but thought it was an Op Room thing...never realized the importance of it pre and post op.*" Director, Surgical Nursing.

Building: "*With a plan in hand, we worked with the team to create the new discharge instructions and related documentation ...*" Director SCN.

Evaluating: "*...my role was to validate this new product...I call patients to follow up as to whether they rec'd discharge instructions and if they were understandable to them... we were at the 40th percentile on this core measure; after one month we were at 73rd percentile; and after two months we were at* $90^{th.}$"Performance Improvement Specialist.

Storing: "*We worked with a physician to get the documentation and instructions included in the patient file.*" Director SCN.

Sharing: "*I just spent an hr discussing our pilot and process with a peer, Director of ICU, who thinks we should standardize the discharge instructions process across the system.*" Director CCU.

## 4 Conclusions

The knowledge development system framework provided a useful way to link the evaluation of clinical improvement services to specific hospital quality practices focused on patient care. Analysis of the context and processes related to actual performance and quality improvement processes across a variety of healthcare organizations will provide an excellent grounding for design of information and communications technology support systems. Identification of the current levels and types of knowledge sharing that can be leveraged and more effectively supported with technology across a wide range of hospitals is the subject of future study.

## References

1. Leonard-Barton, D.: Wellsprings of Knowledge. Harvard Business School Press, Boston (1994)
2. Berwick, D.: Developing and Testing Changes in Delivery of Care. Annals of Internal Medicine 128, 651–656 (1998)
3. Donaldson, M., Mohr, J.: Exploring Innovation and Quality Improvement in Health Care Micro Systems: A Cross-Case Analysis. Institute of Medicine, Washington, DC (2001)
4. Fung-Kee-Fung, M., Goubanova, E., Sequeira, K., Abdulla, A., Cook, R., Crossley, C., Langer, B., Smith, A., Stern, H.: Development of Communities of Practice to Facilitate Quality Improvement Initiatives in Surgical Oncology. Quality Management in Health Care 17(2), 174–185 (2008)
5. Senge, P.: The Fifth Discipline: The Art and Practice of the Learning Organization. Doubleday, New York (1990)
6. Nonaka, I., Toyama, R., Hirata, T.: Managing Flow: A Process Theory of the Knowledge-based Firm. Palgrave, New York (2008)

# Building and Using Terminology Services for the European Centre for Disease Prevention and Control

László Balkányi[1], Gergely Héja[2], and Cecilia Silva Perucha[1]

[1] European Centre for Disease Prevention and Control,
Tomtebodavägen 11a, Stockholm, Sweden
[2] Falcon Informatics Ltd., Szentlászlói út 189, Szentendre, Hungary
laszlo.balkanyi@ecdc.europa.eu

**Abstract.** This paper describes the process of building terminology service and using domain ontology as its conceptual backbone for a European Union agency. ECDC, established in 2005, aims at strengthening Europe's defences against infectious diseases, operates a range of information services at the crossroads of different professional domains as e.g. infectious diseases, EU regulation in public health, etc. A domain ontology based vocabulary service and a tool to disseminate its content (a terminology server) was designed and implemented to ensure semantic interoperability among different information system components. Design considerations, standard selection (SKOS, OWL) choosing external references (MeSH, ICD10, SNOMED) and the services offered on the human and machine user interface are presented and lessons learned are explained.

**Keywords:** ECDC, domain ontology, terminology services, semantic interoperability.

## 1 Introduction

### 1.1 Understanding the Problem – Why ECDC Needs Interoperability Tools?

The European Centre for Disease Prevention and Control (ECDC) builds and operates a range of information services at the crossroads of different professional domains as public health, microbiology, EU regulations. Examples are: The European Surveillance System (TESSy), Threat Tracking Tool (TTT), Intranet Document Repository (content services), Expert Directory. As most of ECDC systems are 'one of their kind' and they were developed under a time pressure to become functional as soon as possible, it was inevitable that on layers of logical design, data modeling, content phrasing, user interfacing, etc., the systems started to diverge. From early on ICT services in ECDC, under sever operational strain, achieved a certain homogeneity on the 'bit-ways' level of operating systems, networking, communication and for the mundane office tasks (served by off-the-shelf products) by choosing tools of one software provider. On the other hand, in between the top, presentation layer and bottom, bit-ways layer, in the middle layer of shared services and contents neither the planning nor the design and implementation of systems have been aligned. Following a careful

P. Kostkova (Ed.): eHealth 2009, LNICST 27, pp. 116–123, 2010.

analysis of the emerging discrepancies and multiplication of coupling needs of different systems, a solution was offered to set up respective layers of interoperability tools / methods / project alignment measures that will on the long run ensure seamless flow of information. This paper focuses on a tool aimed at solving the problems of one layer, the semantic interoperability.

### 1.2 Semantic Interoperability and the Selected Tools: Building a Terminology Server and a Domain Ontology

It is not a surprise, that complex, knowledge intensive organizations, like ECDC working at crossroads of different professions will suffer from the inconsistent labeling of the same concepts. A long history of using enterprise wide data vocabularies in medicine [1], even (in a way) the emergence of markup languages [2] themselves are all symptoms of this phenomenon. The ICT tower of Babel happens to happen over and over again in growing organizations. Analyzing the situation at ECDC [3] it became clear, that implementing simply a rigorous data modeling rule set plus introducing obligatory terminology is not an option because of several reasons: (1) Highly trained professional users (at different, although closely related domains) are prone to use their own professional jargon. (2) Users need different 'granularity', different level of precision in different settings. (3) Grouping, typing, classifying concepts are especially prone to the very specific needs of the given function – you can group communicable diseases according their etiology, the symptomatology of the caused disease, the needed public health measures, the vectors, etc. Therefore ECDC needs a very flexible solution that on one hand allows the specialist to use their known terminology in their known context, on the other hand will gently guide these diverse groups toward a unified terminology, allowing different depth of granularity. To achieve these goals two years ago it was decided that a centrally administered terminology service will be provided, a software application will be built. An agency specific ontology should become the conceptual backbone to connect, to cross map the already existing particular term sets and pushing system users, developers toward a more homogeneous, consistent, navigable multi-domain terminology.

## 2 Methods, Tools, Standards

### 2.1 Avoiding Reinventing the Wheel – What Is Out There?

In order to avoid reinventing the wheel, a survey was done focusing on similar organizations.

**Terminologies, classifications:** Among the (literally) several hundreds of existing medical terminologies we mention here only a subset of them, with very different scope and origin. All of them are relevant as sources for a multi-domain terminology in the area of communicable disease prevention and control, in public health:

The CDC VADS (Public Health Information Network Vocabulary Access and Distribution System [4]) is itself a multitude of vocabularies and a service which allows public health partners to browse, search and download concepts. VADS is huge, with a very rich and broad scope of value sets. Concepts inherit the semantics from the coding system associated with the PHIN vocabulary domain in which they are placed – but there is no overarching semantic model behind the value sets.

The WHO is mandated to the production of international classifications on health. The purpose of the WHO family of classifications (WHO-FIC) is to promote appropriate selection of terms for health fields across the world. It consists of the (1) International Classification of Diseases (ICD), (2) the International Classification of Functioning, Disability and Health (ICF) and (3) the International Classification of Health Interventions (ICHI). All of them have a long legacy compromise along different points of views, and are sometimes abused. Their conceptual frames are not being based on information science principles but on health statistics pragmatics [5].

The largest medical knowledge 'body' made available on the WWW is Medline of the US National Library of Medicine. One of its resources is MeSH (Medical Subject Headings), a controlled vocabulary thesaurus. It consists of sets of terms, descriptors in a hierarchical structure that permits searching at various levels of specificity. There are 25,186 descriptors in MeSH 2009. It is very robust, very well documented and supported collection, but the same way as all the previously mentioned conceptual systems, it lacks a scientifically proved semantic model, enabling e.g. machine inference, or 'logical calculability' [6].

A remarkable recent effort overcoming this lack of being not based on the principles of information science and having relevant, reusable medical or health related content at the same time, resulted in a multi domain health ontology development in the medical arena, called the OBO foundry. This is a 'collaborative' experiment. It involves developers of science-based, health or medicine related ontologies. The developers agreed on a set of common principles for ontology development. The goal is creating a suite of orthogonal interoperable reference ontologies in the biomedical domain [7].

Regarding **standards,** CEN TC 251 and ISO TC 215 are the Technical Committees of standardization in medical informatics on the European and World level [8]. Limitations of current ISO and CEN standards on medical terminology are explained here [9]. International Health Terminology Standards Development Organization is the current steward of Systematized Nomenclature of Medicine – Clinical Terms (SNOMED CT) [10]. SNOMED CT is considered to be the most comprehensive, multilingual clinical healthcare terminology in the world.

Health Level Seven (HL7, [11]) is a US accredited Standards Developing Organization (SDO) operating in the healthcare arena. HL7's domain is clinical and administrative data, it develops specifications. It is probably the most widely used messaging standard that enables disparate healthcare applications to exchange key sets of clinical and administrative data. It has to be mentioned, that there are several serious problem areas, like documentation (intelligibility), implementation, quality of internal consistency of the HL7 RIM etc [12].

Obviously medicine is not the only field where multiple disciplines meet and where huge amount of differently organized and classified information has to be efficiently stored, merged, searched, etc. The Worldwide Web itself is the best source for checking these challenges. Widely used and emerging W3C standards (described later in more details) allow a (hopefully) future proof information modeling and storage in data formats that can be used by a wide array of applications, thereby ensuring transparency.

### 2.2 What Methods, Tools and Standards Have Been Chosen?

To ensure proper support and engineering level integration in ECDC, the terminology server application (TS) was built on MS platform, using an MS SQL engine. Nevertheless for the machine level communication the chosen method was to communicate via standard web services using SOAP interface to allow other applications based on any platform to communicate with the TS. The current core TS uses a predefined set of atomic queries, that can be combined, but the next version already under development will allow SPARQL [13] queries.

Terminologies are represented according to the format of an emerging W3C standard, SKOS [14]. SKOS provides a standardized way to represent knowledge organization systems (as e.g. thesauri, taxonomies, classifications and subject heading systems) using (a limited subset of) OWL. SKOS allows information to be passed between applications in an interoperable way. SKOS also allows knowledge organization systems to be used in distributed, decentralized metadata applications.

The ontology, that serves as conceptual backbone and where (other than just hierarchic) relations among concepts are stored, is stored as an Ontology Web Language (OWL) file. The decision to represent the ontology in OWL rather than SKOS is due to some limitations of SKOS in expressing relations and allowing certain operations among terms. OWL, another W3C standard, was designed for use by applications that need to process the content of information instead of just presenting. OWL can be used not only to explicitly represent the meaning of terms in vocabularies but also all kind of relationships between those terms [15]. OWL has more facilities for expressing meaning and semantics than SKOS, and thus OWL goes beyond in its ability to represent machine interpretable content on the Web.

## 3 Results in Building Terminology, Operating the Terminology Server (TS) and Planned Next Steps

**Content:** The TS contains two `content levels`: `value sets` on the bottom level and `ontology` on the top level. On the *`value set` level* there are three collections of terminologies: (1) application specific sets, like variables of the Threat Tracking Tool, (2) common, shared value sets, e.g. Pathogen Organisms and (3) external reference value sets, e.g. ECDC relevant MeSH terms.

For each term we store a preferred label, alternative labels, a valid/obsolete flag, as well as other metadata items. If a term becomes obsolete, there is a relation to its succeeding, new term. Relationships to `parents` and `children` are stored as well - plus a binding relation to a concept in the backbone ontology. This binding relation ensures a `mapping` of a term`s meaning from one value set across the ontology to any other value set.

*On the ontology level* ECDC ontology uses DOLCE [16], a domain neutral top level ontology for top level concepts, and tries to build scientifically correct domain ontology. The ontology currently describes concepts the following domains:

- organism, based on the biological taxonomy. Most organisms are human pathogens, but hosts (e.g. swine) and vectors (e.g. mosquito) are also listed.
- anatomical structures, currently on organ and organ system level. The anatomical model is based on FMA [17].

**Table 1.** Existing and 'under construction' value sets in ECDC Terminology Server

| Terminology set types | Examples | Estimated size (no. of terms) |
|---|---|---|
| Application specific sets: | Threat Tracking Tool (TTT) | ~ 160 |
| | The European Surveillance System (TESSY) | ~ 280 |
| | Web Portal topics | ~ 85 |
| Common, shared sets: | Pathogen Organisms | ~ 4700 |
| | Communicable diseases, conditions, syndromes | ~ 1200 |
| | Public health terms | ~ 1600 |
| | Geo entities (countries, cities, regions) | ~ 40500 |
| | Organizations | ~ 400 |
| | Administrative terms, abbreviations, acronyms | ~ 200 |
| External reference sets: | ICD 10 (relevant subset) | ~ 1200 |
| | MeSH (relevant subset) | ~ 3760 |
| | SNOMED (relevant subset) | ~ 16500 |
| | Sum: | ~ 70600 |

- diseases, in the mandate of ECDC are represented, however the extension to all infectious diseases from ICD10 is under way.
- medical and epidemiological actors, activities and events.

The ontology is intended to be used later for automatic reasoning - consequently it is represented in OWL DL. The tool used for ontology building is Protégé. Because the top-level of the ontology by its nature deals with quite abstract notions, in its original from is not very usable for physicians. For that reason during the upload of the ontology into the TS it is converted to a semantic network that is easier to understand by physicians. Table 2 displays the top-level of this (simplified) semantic network.

**Table 2.** ECDC domain ontology, top level classes and properties

| Generic classes |
|---|
| Activity |
| Anatomy |
| Biological process |
| Collection |
| Disorder |
| Equipment |
| Geopolitical entity |
| Legal regulation |
| Material |
| Organization |
| Organism |
| Personal role |

| Properties | | |
|---|---|---|
| attribute | | |
| relation | | |
| | caused by | causes |
| | has member | member of |
| | has temporal feature | temporal feature of |
| | locative relation | inverse locative relation |
| | partitive relation | inverse partitive relation |
| | produced by | produces |
| | resistance of | resistant to |

**Functions of TS:** The Terminology Server answers queries, using a basic set of them that can be combined for more sophisticated querying. The client applications use this set of queries retrieving terminology information (synonyms, related terms, time line of changes, etc). The terminology server has also a human user interface, allowing experts to browse all the value sets, and the ontology, to navigate in this term 'space' along the defined relations among categories (of value sets) and concepts (of ontology).

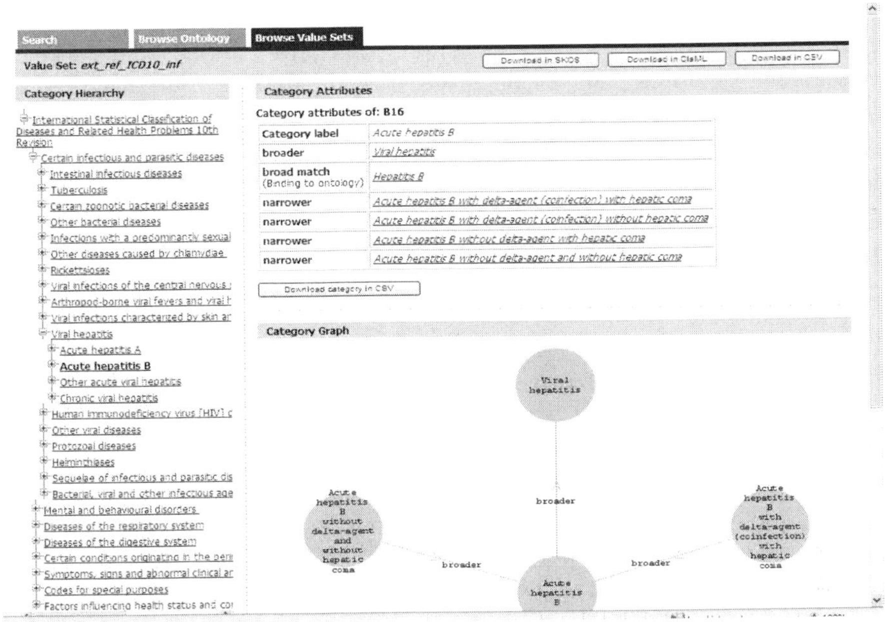

Fig. 1. Screen shot of human user interface of ECDC core terminology server

The extended version that is under development at the publication of this paper will provide enhanced complex query services, query using SPARQL expressions, alert notification to value set administrators, etc. The future plans include Web publication of ECDC terminology services, made available for public health system developers of EU member states, both on the human and the machine interface.

**Table 3.** Services of the core terminology server

| Functionality | Explanations |
|---|---|
| Read operations: | *Domain of information:*<br>• Ontology, concepts and relations;<br>• Value sets and categories.<br>*Operations:*<br>• Get detailed or short description of data elements;<br>• Navigate the data elements according to the retrieved relations;<br>• Search of data elements;<br>• Download the ontology and value sets in OWL, SKOS, ClaML, ... format. |
| Write operations: | *Domain of information:*<br>• Ontology, concepts and relations;<br>• Value sets and categories.<br>*Operations:*<br>• Importing the ontology and value sets (SKOs, OWL)<br>• Creating, modifying and deleting value sets and categories. |
| Retrieval of a certain conceptual element: | E.g.: GetValueSets, GetValueSet, GetCategory, GetOntology, GetConcept and GetRelation); |
| Search for conceptual elements: | With matching the query natural language text, e.g.: SearchValueSet, SearchCategory, SearchConcept |

The functions described in table 3 are / will be used by ECDC client applications to populate certain fields in their interactive forms; as source for (semi)automated meta-data tagging; to enable 'time machine' functionality allowing to use old (obsolete) versions of terms while working with 'old' but still scientifically valuable content. (In epidemiology, data and information is re-used over decades, e.g. data on 'Spanish flu' is reused in studies of understanding current influenza pandemic.)

## 4 Discussion and Conclusions

Why to build and use an in-house terminology service, while there are a number of well managed, comprehensive sources out there? Although these sources are very valuable in providing references and standard approach in designing structure and functions, obviously they will never answer to the specific internal needs of a given organization. We need navigation within our own complex term space with multiple views, answering conflicting granularity needs, clustering along different perspectives specific to ECDC, etc.

The experiences to build up a shared terminology service for ECDC taught us, that such an approach (setting up a standard based, ontology 'enhanced' terminology server) has no serious technical, IT engineering obstacles these days. The existing standards on different levels allow a straightforward, transparent approach, so far the developers of the client systems were able to interpret and use (parse) the messages of the TS without significant problems. Lessons learned:
(1) Both to achieve internal consensus on and to build up shared terminology contents needed significantly more resources than planned at the beginning.
(2) Existing external reference term sets proved to be relatively poor regarding the domain of public health.
(3) Although using SKOS for value sets and OWL for the ontology caused some problems in how to build up the bindings and navigating functionality among the two levels, this has been solved by some restrictions on what OWL constructs could be used.

We think that the chosen approach, based on W3C rather than CEN TC 251 or ISO TC 215 standards, might be future proof and adds practical interoperability. We hope that by publishing the public health and communicable disease related terminology services we will help to fulfill the mandate of ECDC in assisting member states. This work triggered also several interoperability efforts on lower level of system to system messaging and also on higher level of services alignment. Further steps in utilizing the terminology services in building semantic search and knowledge navigation tools will follow.

## References

1. Cimino, J.: Desiderata for controlled medical vocabularies in the twenty-first century. Methods Inf. Med. 37(4-5), 394–403 (1988)
2. Goldfarb, C., Rubinsky, Y.: The SGML handbook. Oxford University Press, Oxford (1990)

3. Balkányi, L.: Terminology services: an example – an example of knowledge management in public health. Euro Surveill. 12(22) (2007)
4. Public Health Information Network Vocabulary Access and Distribution System, http://phinvads.cdc.gov/vads/SearchVocab.action
5. Family of International Classifications, http://www.who.int/classifications/en/
6. Medical Subject Headings, http://www.nlm.nih.gov/pubs/factsheets/mesh.html
7. Open Biomedical Ontologies, http://www.obofoundry.org/
8. CEN TC 251, http://www.cen.eu/CENORM/Sectors/TechnicalCommitteesWorkshops/CENTechnicalCommittees/CENTechnicalCommittees.asp?param=6232&title=CEN%2FTC+251, ISO TC 215, http://www.iso.org/iso/iso_technical_committee?commid=54960
9. Rodrigues, J.M., Kumar, A., Bousquet, C., Trombert, B.: Standards and biomedical terminologies: the CEN TC 251 and ISO TC 215 categorial structures. Stud. Health. Technol. Inform. 136(issue), 857–862 (2008)
10. IHTSD, http://www.ihtsdo.org/
11. Health Level Seven, http://www.hl7.org/
12. HL7, problem areas, http://hl7-watch.blogspot.com/2005/11/list-of-problem-areas.html
13. SPARQL Protocol and RDF Query Language, http://www.w3.org/TR/rdf-sparql-query/
14. Simple Knowledge Organization System, http://www.w3.org/TR/2009/CR-skos-reference-20090317/
15. Ontology Web Language, http://www.w3.org/TR/owl-features/
16. DOLCE: http://www.loa-cnr.it/DOLCE.html
17. Rosse, C., Mejino Jr., J.L.V.: A Reference Ontology for Biomedical Informatics: the Foundational Model of Anatomy. J. Biomed. Inform. 36(6), 478–500 (2003)

# Semantic Description of Health Record Data for Procedural Interoperability

Jan Vejvalka[1], Petr Lesný[1], Tomáš Holeček[3], Kryštof Slabý[1], Hana Krásničanová[1], Adéla Jarolímková[2], and Helena Bouzková[4]

[1] Faculty Hospital Motol and 2nd Faculty of Medicine, Charles University, Prague, Czech Republic
[2] CESNET, z.s.p.o., Prague, Czech Republic
[3] Faculty of Humanity Studies, Charles University, Prague, Czech Republic
[4] National Medical Library, Prague, Czech Republic
{jan.vejvalka,petr.lesny,krystof.slaby}@lfmotol.cuni.cz

**Abstract.** Growing volume of knowledge that needs to be processed and communicated leads to penetration of information and communication technologies (ICT) into biomedicine. Specialized tools for both algorithmic processing and for transport of biomedical data are developed. Proper use of ICT requires a proper computer representation of these data and algorithms. We have analyzed the openEHR archetypes in order to utilize openEHR formatted data in medical grid environments on the MediGrid platform. Both openEHR and MediGrid utilize the phenomenological approach to biomedical data; however the level of constraint placed by both systems on the concepts transported or processed data is different.

**Keywords:** openEHR, archetypes, MediGrid, phenomenology, medical algorithms, health records, semantics, ontologies.

## 1 Introduction

Growing volume of scientific knowledge and resulting specialization in medicine stresses the importance of setting and implementing algorithms and guidelines, which objectify and standardize healthcare processes. Despite the differences in understanding of standardization and algorithmization between the world of medical practice and the world of information and communication technologies (ICT), ICT are often used to support standardization in medicine. Proper use of ICT requires correct computer representation of relevant data and algorithms. Several successful projects aiming at creating or documenting a collection of biomedical algorithms exist (such as MEDAL [5], MedCalc etc.).

OpenEHR [1] is Open Source Electronic Health Record architecture, based on the strict ontology description principles. OpenEHR is based upon archetypes – computable expressions of a domain content model in the form of structured constraint statements, based on reference (information) models of clinical concepts.

P. Kostkova (Ed.): eHealth 2009, LNICST 27, pp. 124–130, 2010.

MediGrid [2] is an open-source algorithm description system which was developed in reaction to clinical requirements for correct biomedical data algorithm processing. It minimizes the possibility of errors caused by most common misuses of clinical data in biomedical algorithms (cf. Table 1) by providing tool to manipulate the domain ontologies of biomedical algorithms.

**Table 1.** Common misuses of biomedical algorithms on clinical data

| Error | Examples |
|---|---|
| Application of good algorithm to wrong data | *Trivial:* entering weight in pounds into the common BMI formula.<br>*Less trivial:* using incorrect formula for body surface area (more than 4 formulae exist, each giving different and in the case of small children even wrong results) for drug dosage. |
| Selection of wrong algorithm | *Trivial:* using standard deviation instead of percentiles in normalization of body weight<br>*Less trivial:* use of body mass index percentile instead of the weight for height ratio percentile for proportionality in early post-pubertal girls, leading to false statements of obesity. |

The purpose of our work presented in this paper was to postulate the principles of MediGRID as a system for semantic-based representation of biomedical knowledge and to compare MediGRID and openEHR. We compared them from the perspective of practical applicability in knowledge domains corresponding to specific areas of medicine, aiming to find possible complementarities in the respective approaches.

## 2 Materials and Methods

MediGrid is being designed to facilitate analysis of semantics automatic processing of semantically well described biomedical data with scientific knowledge represented by semantically well described biomedical computational algorithms, with the means of up-to-date information technologies.

MediGrid approach aims to representation of biomedical data and algorithms as resources, and on sharing these resources in a grid-like environment. The natural way of assigning biomedical algorithms to data (or vice versa) is by comparing their semantic values; therefore MediGrid builds its mechanisms of sharing on capturing and processing of semantic information. Results from previous projects (SMARTIE, Growth 2) have been used as a base for our work: we used existing algorithms to verify our results and we used our experience from SMARTIE to structure semantic information.

The basic assumptions of MediGrid are:

- Data processed by biomedical algorithms are (following the philosophical tradition of phenomenology as formulated by Husserl [3]) indicators[1] that can be transformed into other indicators and grouped into indicator classes by their roles in these transformations.
- Data and algorithms can be shared across conceptual domains if trusted semantic links exist to support such interconnection.

We have performed retrospective analysis of current representations of algorithms and algorithm repositories and formulated several key principles, based on a sound philosophical background, on accepted principles of documentation and scientific communication, on which our approach is based. We have implemented these principles in software and validated their usefulness and effectiveness on the application domain of paediatric auxology.

# 3 Results

## 3.1 Description of Medigrid Entities

The need for extensive review and verification of semantic links when matching data with algorithms, and also the stress on correct procedures in medicine (lege artis) result in practical requirements on design and implementation of a knowledge-processing system based on these theoretical principles:

- Semantic information (meaning for the human user) of both indicator classes and transformations must be explicitly described and readily available for user assessment and validation.
- Semantic information must be bound to the current scientific paradigm and to evidence based medicine through extensive links to published and reviewed works.
- Mechanisms of procedural authority and trust must be implemented to support users' decisions about procedural values of individual components.

We have identified several logical layers, which need to be documented in order to meet the requirements mentioned above:

- Source layer, which contains the description of supporting information: author and cited work.
- Concept layer, which contains semantic description of the two essential categories – transformations (algorithms) and indicator classes (data entering or being exchanged between the algorithms)

[1] "A thing is... properly an indication if and where it in fact serves to indicate something to some thinking being... a common circumstance [is] the fact, that certain objects or states of affairs *of whose reality someone has actual knowledge indicate to him the reality of certain other objects or states of affairs, is the sense that his belief in the reality of the one is experienced ... as motivating a belief or surmise in the reality of the other.*" [Italics by E. H.].

- Implementation layer, which contains information about the specific implementation of transformations and of validations of indicator classes in computer programs.
- Review/trust layer, containing user reviews and trust statements.

In our present implementation, each description on the conceptual layer consists of four basic elements:

- Human semantics: collection of human readable description, which helps the user to understand the meaning of the entity
- Metadata: pieces of computer recognizable data, which can be utilized e.g. to construct user interfaces
- Relations: relations of this entity to other entities
- Classifications: special case of relations that position the entity into external classification and terminology systems (UMLS, MeSH descriptors etc.).

Based on these principles, we have built a proof-of-concept implementation of a tool for semantics-based matching of data and algorithms. Based on the concepts described above, we dissected significant parts of the software "Compendium of Paediatric Auxology 2005" into a set of MediGrid indicator classes, transformations and their implementations. The software (MediGrid tool and the Paediatric Auxology example) is available at SourceForge; a pilot implementation on a web server is being used for testing and further development of the software.

In certain aspects, the Medigrid approach is very close to openEHR; we further analyzed similarities and differences.

## 3.2 MediGRID and OpenEHR

Both openEHR and MediGrid share some common features; they are based on the same principles, e.g. the "two-level" modeling of information, and both of them utilize phenomenological approach [3] in the description of the clinical data. The observable or computable clinical data are described as *Indicators* in MediGrid or *Observations* in openEHR; the grouping of the data in *Contexts* (MediGrid) is paralleled by the archetypes *Cluster* and/or *Folder* in openEHR. Due to this intrinsic compatibility, we are able to utilize the MediGrid methodology to process openEHR-encoded data. From the strict point of view, MediGrid entity models and entities fall into the archetype description, as defined in [4].

Carrier of the semantic information in the openEHR is the Details section, containing the definitions of use, misuse, copyright statement and the original resource URI. The mapping of the openEHR concepts to the entities required to fully semantically describe algorithms (utilized within the framework of MediGrid) is contained in Table 2. In order to utilize biomedical data bound to openEHR archetypes, the semantic constraints of both systems should match. The example of the concept of Body height, presented in openEHR and MediGrid (in two knowledge domains) is shown in Table 3. Similar tables could be constructed for other archetypes existing both in the openEHR and in MediGrid.

**Table 2.** Key elements of openEHR and MediGrid

| Element | MediGrid | openEHR |
|---|---|---|
| User (reference to the author, reviewer, user of the software, implementer etc.) | User | Text encoded within each archetype |
| Existing (e.g. published) data reference available for peer review | Reference | Text encoded within each archetype |
| Observable (or computed) clinical data model (e.g. the body weight or body height) | Indicator class | Observation archetype |
| Hierarchical structuring mechanism (e.g. population ▶ patient ▶ exam ▶ E.N.T. exam) | Context | Cluster or Folder archetype |
| Data processing algorithm or procedure (for automatic or human application, e.g. BMI) | Transformation | Not implemented |
| Description of the algorithm's specific implementation | Implementation | Not implemented |
| Description of the accepted data | Validation | Encoded in the definition in each archetype |
| Mechanism for peer reviewing and trust statements | Review | Missing. Not required? |

**Table 3.** Comparison of the Body height constraints

| Declaration | openEHR | MediGrid: pneumology | MediGrid: anthropometry |
|---|---|---|---|
| Purpose | For recording the height or length of a person at any point in time, and in addition tracking growth and loss of height over time. | For application in the pneumological algorithms | For application in the anthropometrical algorithms, including the tracking of growth and growth dynamics |
| Constraints | Position – lying or standing<br>Not to be used for growth velocity<br>Difference from birth length<br>Not estimated or adjusted | Position: standing<br>Difference from birth length<br>Not estimated or adjusted | Body height is defined as only in standing position (Difference from birth length)<br>[Designed for use in calculations of growth velocity] |
| Measurement | The length of the body from crown of head to sole of foot. | Any allowed | Precise measuring method described |

From the 166 currently existing archetypes described in openEHR and 94 entities of MediGrid implemented in the Faculty Hospital in Prague Motol, there are 10 concepts occuring in both systems. With the only exception (patient sex, which is required by e.g. the anthropometrical or pneumology algorithms and which is moved from the openEHR archetypes into the demographical section of the openEHR record), the data described in openEHR archetypes could be presented as MediGrid indicators and processed by the algorithms from the growing MediGrid library.

Another important difference is in the process of version checking, where the openEHR utilizes the versioning with backwards compatibility, whereas the MediGrid utilizes the quality control procedures and changes propagation.

# 4 Discussion

## 4.1 Indicator Ontology

The term biomedical ontology often refers to structuring the biomedical domain knowledge, e.g. disease taxonomies, medical procedures, anatomical terms, in a wide variety of medical terminologies, thesauri and classification systems. Besides various ontologies that describe the domain of anatomy, the most known of these systems are ICD, SNOMED-CT, GO, MeSH and UMLS, consisting usually of a thesaurus of biomedical terms or concepts on one side and a set of relationships between them on the other side. The basic hierarchical link between the thesaurus elements in most of these systems is the "is_a" relationship. If one element "is_a" another element then the first element is more specific in meaning than the second element. Other widely used relationships are "part_of", "result_of", "consist_of" or "associated_with". These relationships are perfectly suitable for describing basic semantic, functional or topologic structures. For more complex relations between concepts, as e.g. those represented in the domain of paediatric auxology, the ontology of indicators, indicator classes and transformations is much more productive: it allows to define and describe the key concepts and relations that constitute the domain, irrespective of their position in any general terminology / classification / ontology system. At the same time, if desired, relations to these "external" systems can be introduced in order to support mapping of concepts between domains.

## 4.2 Archetype Data as Indicators

Introducing the concept of indicators, indicator classes and indicator transformations, available for user assessment and subject to user decisions, into the archetype model of domain representation may contribute to interoperability of data and knowledge between different domains (represented by different archetypes). In fact, the use of explicitly documented indicator classes and transformations as archetype elements may help to maintain the archetypes as generic domain knowledge representation while allowing to process specific data (originating in different domains) and to use specific parts of that knowledge.

# 5 Conclusions

The need for proper support of traditional ways of medical information handling leads to specific requirements on the ways biomedical data and algorithms and their semantics are documented. Theoretical principles for implementation of a semantic-based tool for representation and application of biomedical knowledge have been postulated and verified by a non-trivial practical implementation. These principles may serve as a further step towards empowerment of users for better control over the knowledge that is represented in the ICT tools they are using.

Algorithms described in MediGrid notation can be directly used on data compliant with openEHR specifications (taking into account some specific issues). In such a way, data storage and transport, which is the main purpose of EHR systems, can be complemented with processing of the data in biomedical algorithms.

## Acknowledgements

Supported by Czech research projects MediGrid, 1ET202090537 and by VZ FNM, MZO 00064203.

## References

1. Fernandez-Breis, J.T., Menarguez-Tortosa, M., Martinez-Costa, C., Fernandez-Breis, E., Herrero-Sempere, J., Moner, D., Sanchez, J., Valencia-Garcia, R., Robles, M.: A Semantic Web-based System for Managing Clinical Archetypes. In: Conf. Proc. IEEE Eng. Med. Biol. Soc. 2008, pp. 1482–1485 (2008)
2. Vejvalka, J., Lesny, P., Holecek, T., Slaby, K., Jarolimkova, A., Bouzkova, H.: MediGrid - Facilitating Semantic-Based processing of Biomedical Data and Knowledge. In: Karopka, T., Correia, R.J. (eds.) Open Source in European Health Care:The Time is Ripe, pp. 18–21. INSTICC Press, Porto (2009)
3. Husserl, E.: Logical Investigations, Investigation I (translated by J. N. Finday), p. 184. Routledge, London (2001)
4. Beale, T.: Archetypes and the EHR. Stud. Health Technol. Inform. 96, 238–244 (2003)
5. The medical algorithms project, `http://www.medal.org`
6. Stroud, S.D., Erkel, E.A., Smith, C.A.: The use of personal digital assistants by nurse practitioner students and faculty. J. Am. Acad. Nurse Pract., 67–75 (2005)

# A Lexical-Ontological Resource for Consumer Healthcare*

Elena Cardillo, Luciano Serafini, and Andrei Tamilin

FBK-IRST, Via Sommarive 18, 38050 Povo (Trento), Italy
{cardillo,serafini,tamilin}@fbk.eu

**Abstract.** In Consumer Healthcare Informatics it is still difficult for laypeople to find, understand and act on health information, due to the persistent communication gap between specialized medical terminology and that used by healthcare consumers. Furthermore, existing clinically-oriented terminologies cannot provide sufficient support when integrated into consumer-oriented applications, so there is a need to create consumer-friendly terminologies reflecting the different ways healthcare consumers express and think about health topics. Following this direction, this work suggests a way to support the design of an ontology-based system that mitigates this gap, using knowledge engineering and semantic web technologies. The system is based on the development of a consumer-oriented medical terminology that will be integrated with other medical domain ontologies and terminologies into a medical ontology repository. This will support consumer-oriented healthcare systems, such as Personal Health Records, by providing many knowledge services to help users in accessing and managing their healthcare data.

**Keywords:** Medical Vocabulary Acquisition, Consumer-oriented Terminologies, Healthcare Ontologies.

## 1 Introduction

With the advent of the Social Web and Healthcare Informatics technologies, we can recognize that a linguistic and semantic discrepancy still exists between specialized medical terminology used by healthcare providers or professionals, and the so called "lay" medical terminology used by patients and healthcare consumers in general. The medical communication gap became more evident when consumers started to play an active role in healthcare information access. In fact, they have become more responsible for their personal healthcare data, exploring health-related information sources on their own, consulting decision-support healthcare sites on the web, and using patient-oriented healthcare systems, which allow them to directly read and interpret clinical notes or test results. To help consumers fill this gap, the challenge is to sort out the different ways consumers communicate within distinct discourse groups and map the common,

* This work is supported by the TreC Project, funded by the Province of Trento.

P. Kostkova (Ed.): eHealth 2009, LNICST 27, pp. 131–138, 2010.

shared expressions and contexts to the more constrained, specialized language of healthcare professionals. In particular, medical Knowledge Integration in healthcare systems is facilitated by the use of Semantic Web technologies, helping consumers during their access to healthcare information and improving the exchange of their personal clinical data. Though much effort has been spent on the creation of these medical resources, used above all to help physicians in filling in Electronic Health Records, facilitating the process of codification of symptoms, diagnoses and diseases, there is little work based on the use of consumer-oriented medical terminology, moreover most of existing studies focused only on English.

Given this scenario, we want to propose a methodology for the creation of a lexical-ontological resource oriented to healthcare consumers in the Italian context, and its integration with a coherent semantic medical resource, useful both for professionals and for consumers. Such a resource can be used in healthcare systems, like Personal Health Records (PHRs), as to help consumers during the process of querying and accessing healthcare information, so as to bridge the communication gap. In the present work we focus in particular on the use of a hybrid methodology for the acquisition of consumer-oriented "lay" terminology for expressing medical concepts such as symptoms, diseases, anatomical concepts, for the consequent creation of a Consumer Medical Vocabulary for Italian, which can be used to translate technical language with lay terminology and vice-versa.

## 2 Medical Terminologies and Ontologies

Over the last two decades research on Medical Terminologies has become a popular topic and the standardization efforts have established a number of terminologies and classification systems, such as SNOMED International[1] or ICD-10 (International Classification of Diseases)[2], as well as conversion mappings between them to help medical professionals in managing and codifying their patients health care data. They concern with *"the meaning, expression, and use of concepts in statements in the medical records or other clinical information systems"* [6]. Having all these medical terminologies interoperability has become a significant problem. Content, structure, completeness, detail, cross-mapping, taxonomy, and definitions vary between existing vocabularies. During the last few years, thanks to the Semantic Web perspective, a set of new methodologies and tools were generated for improving healthcare systems, in particular translating medical terminologies into more formal representations using ontology methodologies and languages (e.g., the formalization of SNOMED CT [7]).

Much effort has also been spent for the creation of new Biomedical Ontologies, such as the Foundational Model Anatomy (FMA) [5], and GALEN into OWL. Ontologies in the medical domain provide an opportunity to leverage the capabilities of OWL semantics and tools to build formal, sound and consistent medical terminologies, and to provide a standard web accessible medium

[1] `http://www.ihtsdo.org/snomed-ct/`

[2] `http://www.who.int/classifications/icd/en/`

for interoperability, access and reuse. Given the presence of all these medical ontologies, two other important issues have to be taken into account: Ontology Mapping, to show how concepts of one ontology are semantically related to concepts of another ontology; and Ontology Integration, which allows access to multiple heterogeneous ontologies[3].

Despite these advantages, the vocabulary problem continues to plague health professionals and their information systems, and especially laypeople, who are the most damaged by the increased communication gap. To respond this consumer needs, during the last few years, many researchers have labored over the creation of lexical resources that reflect the way healthcare consumers express and think about health topics. One of the largest initiatives in this direction is the Consumer Health Vocabulary Initiative[4], by Q. Zeng and colleagues at Harvard Medical School, resulted in the creation of the Open Access Collaborative Consumer Health Vocabulary (OAC CHV) for English. It includes lay medical terms and synonyms connected to their corresponding technical concepts in the UMLS Metathesaurus. They combined corpus-based text analysis with a human review approach, including the identification of consumer forms for "standard" health-related concepts. An overview of all these studies can be found in Keselman *et al.* [3].

It is important to stress that there are only few examples of the application as far as these initiatives are concerned. For example, in Kim *et al.* [4] and Zeng *et al.* [9] we find an attempt to face syntactic and semantic issues in the effort to improve PHRs readability, using the CHV to map their content. On the other hand, Rosembloom *et al.* [8] developed a clinical interface terminology, a systematic collection of healthcare-related phrases (terms) to support clinicians' entries of patient-related information into computer programs such as clinical "note capture" and decision support tools, facilitating display of computer-stored patient information to clinician-users as simple human-readable texts.

## 3 Approach

The global approach followed for this research activity is divided in two macro phases. The first one includes the creation of a Consumer Medical Vocabulary (CMV) for Italian, for collecting common medical expressions and terms used by Italian speakers. The second one focuses on the formal representation of some relevant medical terminologies, which will be integrated with the CMV, and the development of a Medical Ontology Repository (MORe) in which all these ontologies and terminologies will be integrated. This activity can be further characterized by the following tasks:

- Knowledge Acquisition/Terminology Extraction. Use of elicitation techniques to acquire all the lay terms, words, and expressions, used by laypeople to indicate specific medical concepts.

[3] `http://www.obofoundry.org`
[4] `http://www.consumerhealthvocab.org`

- Creation of the Italian Consumer Medical Vocabulary (CMV). Selection of all the lay terms extracted that have been identified as good representatives for medical concepts (considered as synonyms after clinical review performed by physicians), and consequent mapping analysis to a standard medical terminology.
- Formalization in terms of OWL. Medical terminologies such as ICD10 and ICPC2 will be formalized into OWL ontologies, and then integrated with the CMV and some existing medical ontologies, relevant for our aims, to guarantee semantic interoperability.
- Creation of a Medical Ontology Repository (MORe) and implementation of Knowledge Services. Some relevant resources will be integrated into MORe, an ontology collection, that will be extended with a set of basic reasoning services to support the implementation of semantic based patient healthcare applications.

In this work we focus on the task of acquisition of consumer-oriented knowledge about a specific subset of healthcare domain, and the creation of the consumer-oriented medical vocabulary for Italian. Concerning the task of Formalization of medical terminologies we refer the reader, for the details of the applied methodology and preliminary results, to Cardillo *et al.* [2]. There we present the formalization of two medical classification systems, ICPC2 and ICD10, into OWL ontologies. In the same work we describe also the construction of a well-founded and medically sound mapping model between the two ontologies by means of the formalization in terms of OWL axioms of the existing clinical mappings, and validation of its coherence using Semantic Web techniques.

### 3.1 Knowledge Acquisition Task

We used a hybrid methodology for the identification of "lay" terms, words, and expressions used by Italian speakers to indicate Symptoms, Diseases, and Anatomical Concepts. Three different target groups were considered: First Aid patients subjected to a Triage Process; a community of Researchers and PhD students with a good level of healthcare literacy, and finally a group of elderly people with a modest background and low level of healthcare literacy. This methodology consisted of the following steps:

1. Application of three different Elicitation Techniques to the mentioned groups;
2. Automatic Term Extraction and analysis of acquired knowledge by means of a Text Processing tool;
3. Clinical review of extracted terms and manual mapping to a standard medical terminology (ICPC2), performed by physicians;
4. Evaluation of results in order to find candidate terms to be included in the Consumer-oriented Medical Vocabulary.

**Wiki-based Acquisition.** The first method is based on the use of a Semantic Media Wiki system, an easy to use collaborative tool, allowing users to create and link, in a structured and collaborative manner, wiki pages on a certain domain of knowledge. Using our online *eHealthWiki* system[5], users created wiki pages for describing symptoms and diseases, using "lay" terminology, specifying in particular the corresponding anatomical categorization, the definition and possible synonyms. The system has been evaluated over a sample of 32 people: researchers, PhD students and administrative staff of our research institute (18 females, 14 males, between 25 and 56 years old). In one month, we collected 225 wiki pages, 106 for symptoms and 119 for diseases, and a total of 139 synonyms for the inserted terms. Users were reluctant to the collaborative functionality of this system, which allows modifying concepts added by others.

**Nurse-assisted Acquisition**. The second method involved nurses of a First Aid Unit[6] as a figure of mediation for the acquisition of terminology about patient symptoms and complaints, helping them to express their problems using the classical subjective examination performed during the Triage Process, which aims to prioritize patients based on the severity of their condition. This method involved 10 nurses, around 60 patients per day and a total of 2.000 Triage Records registered in one month. During this period nurses acquired the principal problems (symptoms and complaints) expressed by their patients using "lay" terminology and inserted them in the Triage Record together with the corresponding medical concepts used for codifying patient data.

**Focus-Group Acquisition**. The method consisted in merging the following elicitation techniques: Focus Group, Concepts Sorting, and Board Games, in order to allow interaction and sharing situations to improve the process of acquisition. The target was a community of 32 elderly people in a Seniors Club, between 65 and 83 year old. We used group activities to acquire, even in this case, lay terms and expressions for symptoms, diseases and anatomy. About 160 medical terms were collected. Then all the terms were analyzed together with other groups, creating discussions, exchanging opinions on terms definitions, synonyms, and recording preferences and shared knowledge. At the end, all participants gave preferences for choosing the right body system categorization (digestive, neurological, musculoskeletal, lymphatic, endocrine, etc.) of each of the written concept.

### 3.2 Term Extraction and Mapping Analysis

The three sets of collected data were further processed and analyzed, to detect candidate consumer-oriented terms, with Text-2-Knowledge tool (T2K) [1]. This tool allowed us to automatically extract terminology from the data sets, to perform many text processing techniques, and to calculate statistics on the extracted data such as term frequency. In spite of the advantages of the automatic extraction process, allowing for extraction of many compound terms, such a procedure has demonstrated that a large amount of terms, certainly representative of consumer medical terminology, were not automatically extracted, since, due

[5] `http://ehealthwiki.fbk.eu`

[6] `http://www.apss.tn.it/Public/ddw.aspx?n=26808`

to the quantitative limits of the corpus dimensions, their occurrence was inferior with respect to the predefined threshold value. Consequently, we performed an additional manual extraction to take into account such rare terms, usually mentioned by a single participant.

Extracted terms were reviewed by two physicians to find incongruences in categorization and synonymy. For instance, "Giramento di Testa" (*Dizziness*) was categorized as Cardiovascular problem instead of Neurological. Physicians have been also asked to map a term/medical concept pair by using a professional health classification system, the above mentioned ICPC2. It addresses fundamental parts of healthcare process: it is used in particular by general practitioners for encoding symptoms and diagnosis. We identified five different types of relations between consumer terms and ICPC2 medical concepts:

- Exact mapping between the pairs; this occurs when the term used by a lay person can be found in ICPC2 rubrics and both terms correspond to the same concept. For example, the lay term "Febbre" (*Fever*) would map to a ICPC2 "Febbre" term, and both will be rooted to the same concept.
- Related mapping; it involves lay synonyms and occurs when the lay term does not exist in the professional vocabulary, but corresponds to a professional term that denotes the same (or closely related) concept. E.g., lay term "Sangue dal Naso" (*Nosebleed*) corresponds to "Epistassi" (*Epistaxis*).
- Hyponymy relation; this occurs when a lay term can be considered as term of inclusion of a ICPC2 concept. E.g., lay term "Abbassamento della Voce" (*Absence of Voice*) is included in the more general ICPC2 concept "Sintomo o disturbo della voce" (*Voice Symptom/Complaint*).
- Hyperonymy relation; in this case the lay term is more general than one or more ICPC2 concepts, so it can be considered as its/their hyperonym. E.g., the term "Bronchite" (*Bronchitis*) is broader than "Bronchite Acuta/ Bronchiolite" (*Acute Bronchitis/ Bronchiolitis*) e "Bronchite Cronica" (*Chronic Bronchitis*) ICPC2 concepts.
- Not mapped; those lay terms that cannot be mapped to the professional vocabulary. These can be legitimate health terms, the omission of which reflects real gaps in existing professional vocabularies; or they can represent unique concepts reflecting lay models of health and disease. E.g., the lay term "Mal di mare" (*Seasickness*).

## 4 First Results Evaluation

We were able to acquire a variegated consumer-oriented terminology and to perform an interesting terminological and conceptual analysis. By means of the term extraction process, from 225 Wiki pages, we were able to extract a total of 962 medical terms. We found a total of 173 Exact Mappings, 80 Related Mappings, 94 Hyperonyms, 51 Hypomyms and, finally, 186 Not Mapped ICPC2 concepts. Most of the exact mappings with ICPC2 are related to anatomical concepts, and many synonyms were found for symptoms. Concerning the Nurse-assisted data set, from 2.000 Triage records we extracted a total of 2389 terms, but about half of these terms were considered irrelevant for our evaluation, so mapping was provided only for 1108 terms. Here we can highlight the high presence of lay terms used for expressing symptoms with exact mappings to ICPC2 (134 on a total of 240 exact mappings), but also many synonyms in lay terminology for ICPC2 concepts (386 Related Mappings). Finally, 321 medical terms were extracted by the transcription of the Focus Group/Game activity (third data set). Here all the symptoms extracted (79 terms) had a corresponding medical concept in ICPC2 terminology (35 Exact Mappings and 44 Related Mappings).

Table 1 below compares the three data sets together and shows that the most profitable method for acquiring consumer-oriented medical terminology was the one assisted by Nurses. Also Wiki-based method, even if not exploited for the collaborative characteristic, has demonstrated good qualitative and quantitative results. Results concerning mapping to ICPC2 can be considered, because 2/3 of the terms extracted are covered by ICPC2 terminology. Comparing the three sets, the overlap is only of 60 relevant consumer medical terms. The overlap with ICPC2 is about 508 medical concepts on a total of 706 ICPC2 concepts. This means that all the other mapped terms can be considered synonyms or quasi synonyms of the ICPC2 concepts. The large number of not mapped terms and the low overlap between the three sets of extracted terms demonstrate that we extracted a very variegated range of medical terms, many compound terms and expressions, which can be representative for the corresponding technical ones in standard terminology, and which can be used as candidate for the construction of our Consumer-oriented Medical Vocabulary for Italian.

**Table 1.** Mapping Results

| Sources | Total Terms | Mapped | Not Mapped |
|---|---|---|---|
| Wiki-based | 962 | 398 | 186 |
| Nurse-assisted | 2389 | 726 | 382 |
| Focus-Group | 321 | 231 | 12 |
| Total | 3662 | 1355 | 580 |

After the task of mapping analysis and the evaluation of the first results, the extracted "lay" terms considered as good synonyms for the ICPC2 symptoms and diseases have been added to the ICPC2 ontology to integrate it with the consumer-oriented terminology.

## 5 Concluding Remarks

In this paper we proposed the creation of a lexical-ontological resource for healthcare consumers that would help to fill in the linguistic communication gap between specialized and "lay" terminology. Such a resource can be used in consumer-oriented healthcare systems in order to help users in accessing to and managing of their healthcare data. We have presented preliminary results for the task of consumer-oriented terminology acquisition, on the basis of statistical and mapping analysis, which helped us to find overlaps between extracted "lay" terms and specialized medical concepts in the ICPC2 terminology. Our methodology showed encouraging results, because it allowed us to acquire many consumer-oriented terms; a low overlap with ICPC2 and a high number of related mappings (mainly synonyms) to the referent medical terminology. To improve the results of the acquisition task and to extract more variegated consumer-oriented terminology, not related to the regional context, we are analyzing written corpora, which include forum postings of an Italian medical website for asking questions to on-line doctors [7]. This will allow extending our sample and cover a wider range of ages, people with different background and consequently different levels of health literacy. This task will be very interesting for comparing results with that came out from the previous elicitation methods, both in quantitative and qualitative terms.

## References


1. Bartolini, R., Lenci, A., Marchi, S., Montemagni, S., Pirrelli, V.: Text-2-knowledge: Acquisizione semi-automatica di ontologie per l'indicizzazione semantica di documenti. Technical Report for the PEKITA Project, ILC. Pisa, 23 (2005)
2. Cardillo, E., Eccher, C., Tamilin, A., Serafini, L.: Logical Analysis of Mappings between Medical Classification Systems. In: Proc. of the 13th Int. Conference on Artificial Intelligence: Methodology, Systems, and Applications, pp. 311–321 (2008)
3. Keselman, A., Logan, R., Smith, C.A., Leroy, G., Zeng, Q.: Developing Informatics Tools and Strategies for Consumer-centered Health Communication. Journal of American Medical Informatics Association 14(4), 473–483 (2008)
4. Kim, H., Zeng, Q., Goryachev, S., Keselman, A., Slaughter, L., Smith, C.A.: Text Characteristics of Clinical Reports and Their Implications for the Readability of Personal Health Records. In: Proc. of the 12th World Congress on Health (Medical) Informatics, MEDINFO 2007, pp. 1117–1121 (2007)
5. Noy, N.F., Rubin, D.L.: Translating the Foundational Model of Anatomy into OWL, in Web Semantics: Science. Services and Agents on the World Wide Web 6(2), 133–136 (2008)
6. Rector, A.: Clinical Terminology: Why is it so hard? Methods of Information in Medicine 38(4), 239–252 (1999)
7. Rector, A., Brandt, S.: Why do it the hard way? The case for an expressive description logic for SNOMED. Journal of American Medical Informatics Association 15(6), 744–751 (2008)
8. Rosembloom, T.S., Miller, R.A., Johnson, K.B., Elkin, P.L., Brown, H.S.: Interface Terminologies: Facilitating Direct Entry of Clinical Data into Electronic Health Record Systems. Journal of American Medical Informatics Association 13(3), 277–287 (2006)
9. Zeng, Q., Goryachev, S., Keselman, A., Rosendale, D.: Making Text in Electronic Health Records Comprehensible to Consumers: A Prototype Translator. In: Proc. of the 31st American Medical Informatics Association's Annual Symposium, AMIA 2007, pp. 846–850 (2007)


[7] `http://medicitalia.it`

# An Ontology of Therapies

Claudio Eccher[1,*], Antonella Ferro[2], and Domenico M. Pisanelli[3]

[1] FBK-irst, Via Sommarive 18, 38050 Povo, Trento, Italy
[2] Medical Oncology Unit, S.Chiara Hospital, Trento, Italy
[3] CNR-Institute of Cognitive Science and Technologies, Rome, Italy

**Abstract.** Ontologies are the essential glue to build interoperable systems and the talk of the day in the medical community. In this paper we present the ontology of medical therapies developed in the course of the Oncocure project, aimed at building a guideline based decision support integrated with a legacy Electronic Patient Record (EPR). The therapy ontology is based upon the DOLCE top level ontology. It is our opinion that our ontology, besides constituting a model capturing the precise meaning of therapy-related concepts, can serve for several practical purposes: interfacing automatic support systems with a legacy EPR, allowing the automatic data analysis, and controlling possible medical errors made during EPR data input.

**Keywords:** ontology, therapy, EPR, NCI thesaurus.

## 1 Introduction

Medicine is a very complex domain from the point of view of modeling and representing intended meaning. In such a discipline we find different activity domains (e.g. clinical vs. administrative knowledge), different scientific granularities (e.g. molecular vs. organic detail), different user requirements for the same service (e.g. physician-oriented vs. patient-oriented views), and ambiguous terminology (polysemy).

Many people today acknowledge that ontologies may help building better and more interoperable information systems, also in medicine [1]. On the other hand, many others are skeptical about the real impact that ontologies - apart from the academic world - may have on the design and maintenance of working information systems.

Ontologies are the talk of the day in the medical informatics community. Their relevant role in the design and implementation of information systems in health care is now widely acknowledged. Ontologies are nowadays considered as the basic infrastructure for achieving semantic interoperability [2]. This hinges on the possibility to use shared vocabularies for describing resource content and capabilities, whose semantics is described in an unambiguous and machine-processable form. Describing this semantics, i.e. what is sometimes called the intended meaning of vocabulary terms, is exactly the job that ontologies do for enabling semantic interoperability.

[*] Corresponding author.

P. Kostkova (Ed.): eHealth 2009, LNICST 27, pp. 139–146, 2010.

Starting from the necessities posed by the design of a breast-cancer treatment decision support system in the course of the Oncocure project, in this paper, we propose an axiomatic ontology for medical therapies, with particular attention to oncologic therapies.

## 2 The Oncocure Project

FBK and S.Chiara Hospital already developed the web-based oncologic Electronic Patient Record (EPR) in strong collaboration with the end users [4]. A considerable effort was made to codify as many data as possible in order to allow their reuse. Initially deployed in 2000 in the Medical Oncology Unit of the S. Chiara Hospital of Trento (Northern Italy), the EPR was subsequently shared with the Radiant Therapy Unit and the Internal Medicine wards of several peripheral hospitals of our province, for enabling the shared management of cancer patients. By now, the EPR stores more than 12,000 cases; breast cancer is by far the most common disease, amounting to about 4,000 cases. The Oncocure project, started in April 2007, intends to design and develop a prescriptive guideline-based CDSS for giving active support at important decisional steps of the oncologic care process, through the execution of the Asbru-encoded protocols of pharmacological therapies for breast cancer. The project aims at integrating the Asbru interpreter with the database and the graphical user interface (GUI) of the EPR, in order to recommend to the user the most appropriate therapeutic strategy in the presence of the specific disease and patient conditions.

Cancer protocols contain a lot of implicit knowledge, especially regarding cancer therapies, for which the name of asset of them implies different intents, temporal relations and cancer phase (e.g., adjuvant, palliative, etc.), different kinds (medical therapy, radiation therapy), different classes of drugs (hormone therapy, chemotherapy). The development of an ontology of therapies can facilitate the interoperability of the two systems by giving a formal definitions of the concepts and their relations regarding the therapy concepts mentioned in cancer guidelines.

## 3 Which Ontologies for Medicine?

What kinds of ontologies do we need? This is still an open issue. In most practical applications, ontologies appear as simple taxonomic structures of primitive or composite terms together with associated definitions. These are the so-called lightweight ontologies, used to represent semantic relationships among terms in order to facilitate content-based access to data produced by a given community. In this case, the intended meaning of primitive terms is more or less known in advance by the members of such community. Hence, in this case, the role of ontologies is more that of supporting terminological services (inferences based on relationships among terms, usually just taxonomic relationships) rather than explaining or defining their intended meaning.

On the other hand, however, the need to establishing precise agreements as to the meaning of terms becomes crucial as soon as a community of users evolves, or multicultural and multilingual communities need to exchange data and services.

To capture the precise meaning of a term and removing ambiguities, we need an explicit representation of the so-called ontological commitments related to these terms. A rigorous logical axiomatisation seems to be unavoidable in this case, as it accounts not only for the relationships between terms, but – most importantly – for the formal structure of the domain to be represented. This allows one to use axiomatic ontologies not only to facilitate meaning negotiation among agents, but also to clarify and model the negotiation process itself, and in general the structure of interaction.

We should immediately note that building axiomatic ontologies for these purposes may be extremely hard, both conceptually and computationally. However, this job only needs to be undertaken once, before the interaction process starts.

Axiomatic ontologies come in different forms and can have different levels of generality, but a special relevance is enjoyed by the so-called foundational ontologies, which address very general domains. Foundational ontologies are ultimately devoted to facilitate mutual understanding and inter-operability among people and machines. This includes understanding the reasons for non-interoperability, which may in some cases be much more important than implementing an integrated (but unpredictable and conceptually imperfect) system relying on a generic shared "semantics".

The role and nature of foundational ontologies (and axiomatic ontologies in general) is complementary to that of lightweight ontologies: the latter can be built semi-automatically, e.g. by exploiting machine learning techniques; the former require more painful human labor, which can gain immense benefit from the results and methodologies of disciplines such as philosophy, linguistics, and cognitive science [3].

## 4 The NCI Thesaurus

One of the most comprehensive lightweight ontologies in the cancer domain is the National Cancer Institute thesaurus (NCIT), define by its authors as "*a biomedical vocabulary that provides consistent, unambiguous codes and definitions for concepts used in cancer research*" and which "*exhibits ontology like properties in its construction and use*" [5]. NCIT is available in OWL [6].

Besides the problem related to the principles that should be adopted in good terminology and ontology design, identified by Ceusters et al. [7], however, NCIT suffers of some problems of classification of cancer therapy concepts. After discussions with the oncologist, we identify a number of these problems, which impair the use of the thesaurus as reference ontology for disambiguating the cancer therapy domain. To give an intuition on the nature of these classification problems, we report here some examples.

1) The class *Neoadjuvant_Therapy*, defined as "Preliminary cancer therapy (chemotherapy, radiation therapy, hormone/endocrine therapy, immunotherapy, hyperthermia, etc.) that precedes a necessary second modality of treatment" is a direct child of *Therapeutic_Procedure*. *Adjuvant_Therapy* on the contrary, defined as "Treatment given after the primary treatment to increase the chances of a cure. Adjuvant therapy may include chemotherapy, radiation therapy, hormone therapy, or biological therapy", is a direct child of *Cancer_Treatment*, which in

its turn is direct child of the classes *Therapeutic_Procedure* and *Cancer_Diagnostic_or_ Therapeutic_Procedure*. No explicit axiom is defined to introduce a temporal relation respect to a primary treatment.

2) *Therapeutic_Procedure* and *Cancer_Diagnostic_or_Therapeutic_Procedure* are distinct sibling classes whose difference is not clear.
3) The classes *Bone_Metastases_Treatment* and *Breast_Cancer_Treatment* are siblings of *Adjuvant_Therapy*, in spite of the fact that the latter represents any kind of cancer therapy with adjuvant intent, while *Breast_Cancer_Treatment* should represent a treatment for a specific tumor (adjuvant, neoadjuvant or metastatic) and *Bone_Metastases_Treatment* is a treatment for a specific metastasis site.
4) *Second_Line_Treatment* and *Protocol_Treatment_Arm* are direct children of *Treatment_Regimen* (in its turn direct child of *Therapeutic_Procedure*). Actually, they are very different concepts: the former is a treatment given after the first treatment has failed; the latter is one arm of a trial protocol, which can be applied in any cancer stage. By the way, the classes *First_line_treatment*, *Third_line_treatment*, etc. are not present in the thesaurus. Instead, there is a *First-line_therapy*, defined as "the preferred standard treatment for a particular condition", and direct child of the class *Therapeutic_Procedure*.
5) *Palliative_Surgery* and *Curative_Surgery*, characterized by the intention for which surgery is performed, are siblings of *Ambulatory_Surgical_Procedure*, characterized by the place in which surgery is performed. A class *Unnecessary_Surgical_Procedure*, apparently characterized by the (a posteriori?) assessment of its uselessness, is sibling of the previous ones.
6) The class *Dose*, defined as "The amount of drug administered to a patient or test subject at one time or the total quantity administered" is a direct child of *Treatment_regimen*. Actually, the dose is not a kind of regimen of treatment, but the quantity of a drug that composes a treatment regimen.

## 5 The Ontology Developed

We implemented a Description Logic (DL)-based ontology in OWL [8], which may be edited and browsed by means of the Protégé tool.

Our ontology is based upon the DOLCE top-level ontology [9]. From DOLCE it inherits the basic distinction between "endurants" and "perdurants". Classically, endurants (also called continuants) are characterized as entities that are 'in time', they are 'wholly' present (all their proper parts are present) at any time of their existence [10]. On the other hand, perdurants (also called occurrents) are entities that 'happen in time', they extend in time by accumulating different 'temporal parts', so that, at any time *t* at which they exist, only their temporal parts at *t* are present. For example, the book you are holding now can be considered an endurant because (now) it is wholly present, while "your reading of this book" is a perdurant, because your "reading" of the previous section is not present now.

In the clinical context, we may say that a medical device is an endurant, whereas a specialty-care visit is a perdurant. In the context of this ontology, the description of a

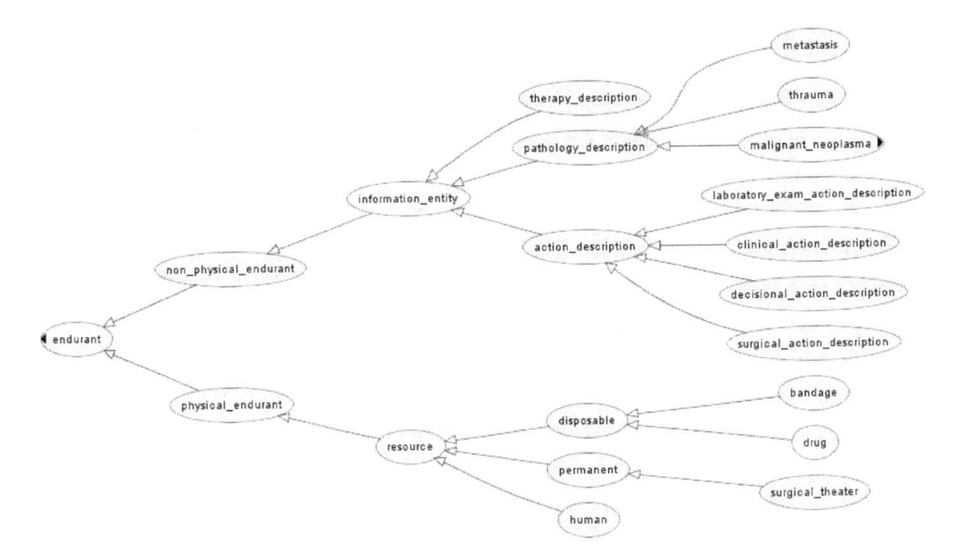


**Fig. 1.** The main endurants of the ontology of therapies

therapy is an endurant and its enactment a perdurant. Figure 1 reports the main endurants defined in the ontology of therapies.

It is natural for a realist based ontology of particulars to take into account physical and tangible objects, but other immaterial and non tangible entities must be considered too. Therefore endurants are distinguished between physical and non-physical ones. The former class is basically devoted to represent resources which may be human, permanent (e.g. surgical theater, lancet) or disposable. Non-physical entities relevant for this ontology are information entity, i.e. therapy description and all the other descriptions useful in the context of a therapy care, like action description and pathology description.

As already mentioned, the most significant perdurants are processes and here we find the different kind of actions (not to be mistaken with their description): clinical, decisional, laboratory-exam and surgical.

What is peculiar to the approach based on DOLCE is the definition of qualities. Qualities are entities such as shape, height, weight which characterize the features of the different items in an ontology. The hierarchy of qualities in the ontology of therapies is reported in figure 2. They are classified according to the method followed (e.g. pharmacological, radiant, and surgical), the pathology involved and their role (e.g. curative vs. palliative, primary vs. non primary, radical vs. non radical).

By attaching several qualities to the same therapy, it is possible to define many possible combinations and avoiding the entanglement of multiple hierarchies. For example, the adjuvant therapy is both a post-operative, and oncological therapy. Rather than establishing two IS_A links with those two entities (multiple hierarchy), we add existential restrictions to adjuvant therapy in order to describe it with the qualities of being post-operative and oncological.

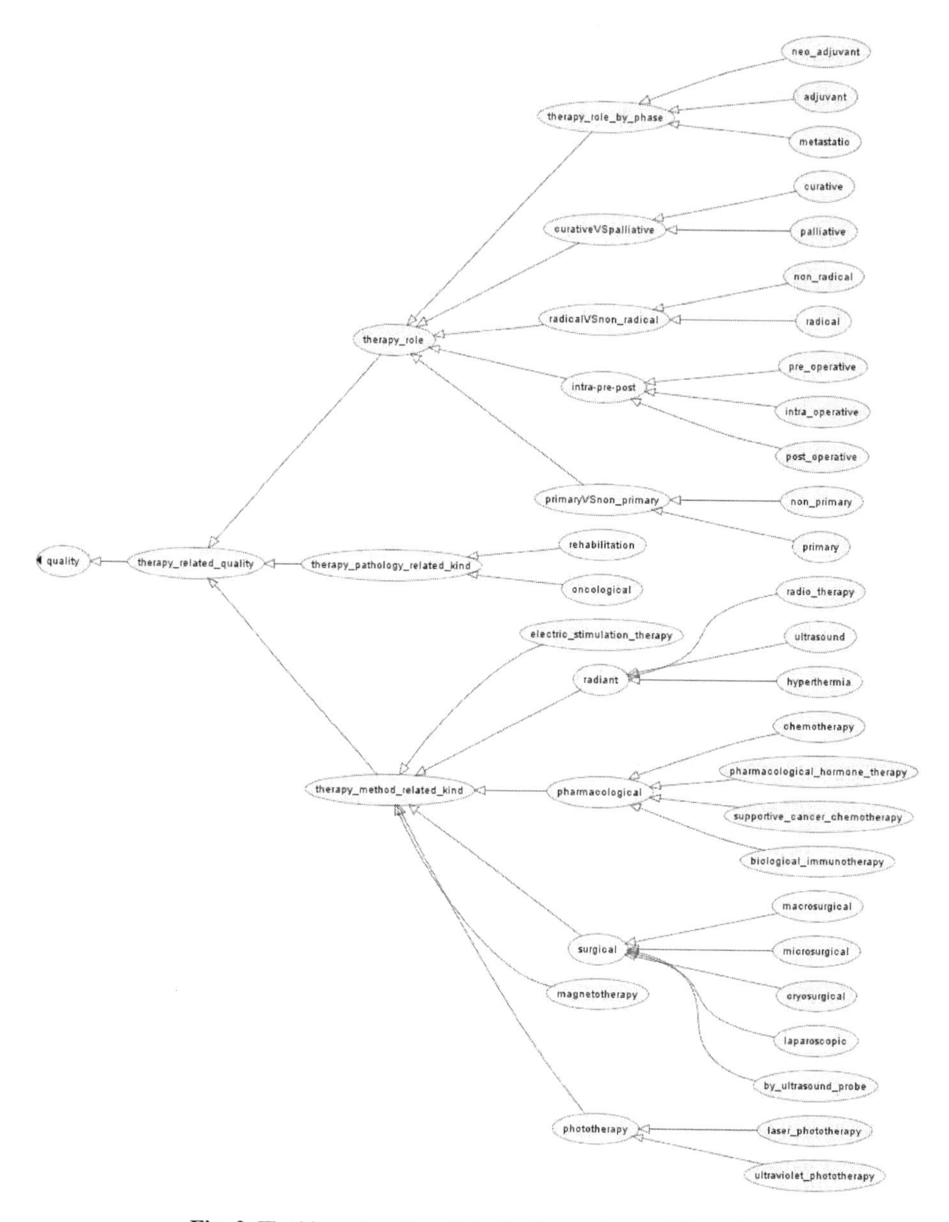


**Fig. 2.** The hierarchy of qualities in the ontology of therapies

## 6 Use of the Ontology

Our axiomatic ontology of medical therapies, besides being a model to capture the precise meaning of a term and removing ambiguities, can be practically used in combination with an electronic Patient Record for several automated tasks. In general, an EPR contain a set of temporally annotated data referring to events: surgery, onset of disease and patient states (e.g., metastases, drug toxicity), prescription of drugs, beginning and ending of therapy administrations. The ontology of medical therapies can allow to label set of raw EPR data with higher level abstraction information defining e.g., role, intent, type of a therapy by various tools supporting the routinely clinician activity.

**Interfacing automatic support systems with legacy EPR.** One of the most relevant research areas in medical informatics concerns the bridging of the gap between a Decision Support System and an Electronic Patient Record. It is often the fact that a DSS model requires parameters relating to the history of therapies administered to the patient in the past. For example, breast cancer guideline can recommend a metastatic treatment with a certain class of drugs (taxanes, aromatase inhibitors) if a drug of the same class was not administered as adjuvant therapy. The ontology can allow a software agent to infer the role of different therapeutic phases in relation to patient and disease states and external events (e.g., neo-adjuvant, adjuvant, metastatic treatments).

**Enabling the automatic data analysis.** Another relevant research topics concerns the automated extraction and visual representation of clinical information EPR for quality control and statistical analysis. The huge amount and complexity of data in EPRs, in fact, renders difficult the 'holistic' view of the patient case. The identification of different therapeutic phases, their classification by using higher level abstractions, and the identification of mutual relations can allow the analysis and exploration of data giving users a comprehensive view of therapy history and therapy effects and, in turn, a deeper understanding of clinical data. Also, the automatic control of the quality of the treatment delivered in a healthcare organization is possible only if the care process can be reconstructed by EPR data and matched with the ideal process recommended by guidelines.

**Controlling the medical errors.** The data input in an EPR can be the source of unreliable data that can influence subsequent decisions and cause serious medical errors. An automatic control of data inputted by the healthcare professionals would be required, able to issuing warnings if the new datum conflicts with stored data. Such a system can exploit the ontology of therapies to connect data representing events, therapies phases (intent, roles) and specific treatments to alert the physician of possible incoherencies in the information introduced; e.g., a drug therapy cannot be adjuvant before surgery or in the presence of a metastatic tumor.

Although the above, not exhaustive, list of examples concerns the possible application of this ontology in the context of the cancer domain, object of the Oncocure project, it is not difficult to foreseen various uses in different domains.

## 7 Conclusions

The interconnection of information processing systems, in order to make optimal use of medical and administrative data, will be the basis for improved care and higher efficiency in future health-care information systems. The relevant role of ontologies in the design and implementation of information systems in health care is now widely acknowledged. The availability of axiomatic ontologies developed according to good design principles and based on foundational ontologies like DOLCE, in fact, can facilitate the interoperability, allowing assigning precise meanings to concepts and solving ambiguities.

Although the NCIT is a useful tool as a reference vocabulary of cancer-related terminologies, which we used as a starting point of our work, we found that the thesaurus is not a good candidate for an ontology of therapies to use in the Oncocure project. Consequently, we developed an ontology of medical therapies, particularly focused on cancer therapies, based on the design principles outlined in Section 5. Without an ontological grounding like that we provide, in fact, the same information may shift its sense according to the context in which it is placed and according to the tacit knowledge of the human agent who specifies its meaning. For example, how do we know that an adjuvant therapy is also a post-operative therapy?

Humans understand the context and have no problems, but computers need ontologies.

## References


1. Pinciroli, F., Pisanelli, D.M.: The unexpected high practical value of medical ontologies. Computers in Biology and Medicine 36(7-8), 669–673 (2006)
2. Pisanelli, D.M. (ed.): Ontologies in Medicine. IOS Press, Amsterdam (2004)
3. Guarino, N. (ed.): Formal Ontology in Information Systems. IOS Press, Amsterdam (1998)
4. Galligioni, E., Berloffa, F., Caffo, O., Tonazzolli, G., Ambrosini, G., Valduga, F., Eccher, C., Ferro, A., Forti, S.: Development and daily use of an electronic oncological patient record for the total management of cancer patients: 7 years' experience. Ann. Oncol. (2008); Epub ahead of print
5. de Coronado, S., Haber, M.W., Sioutos, N., Tuttle, M.S., Wright, L.W.N.: Thesaurus: Using Science-based Terminology to Integrate Cancer Research Results. In: Fieschi, M., Coiera, E., Li, Y.-C.J. (eds.) Proceedings of the 11th World Congress on Medical Informatics, MEDINFO 2004, San Francisco, CA, USA, Sep. 7-11, pp. 33–37. IOS Press, Amsterdam (2004)
6. National Cancer Institute, Office for Communication, Center for Bioinformatics. NCI Terminology browser, `ftp://ftp1.nci.nih.gov/pub/cacore/EVS/` (Last visited May 18, 2009)
7. Ceusters, W., Smith, B., Kumar, A., Dhaen, C.: Mistakes in Medical Ontologies: Where Do They Come From and How Can They Be Detected? In: Pisanelli, D.M. (ed.) Ontologies in Medicine. IOS Press, Amsterdam (2004)
8. W3C. OWL Web Ontology Language Reference. Recommendation (February 10, 2004), `http://www.w3.org/TR/owl-ref/` (Last visited May 18, 2009)
9. Laboratory for Applied Ontology. DOLCE: a Descriptive Ontology for Linguistice and Cognitive Engineering, `http://www.loa-cnr.it/DOLCE.html` (Last visited May 18, 2009)
10. Hawley, K.: How Things Persist. Clarendon Press, Oxford (2001)

# Modelling and Decision Support of Clinical Pathways

Roland Gabriel and Thomas Lux

Chair of Business Informatics, Competence Center eHealth Ruhr,
Universitätsstr. 150, 44801 Bochum, Germany
{roland.gabriel,thomas.lux}@winf.rub.de

**Abstract.** The German health care market is under a rapid rate of change, forcing especially hospitals to provide high-quality services at low costs. Appropriate measures for more effective and efficient service provision are process orientation and decision support by information technology of clinical pathway of a patient. The essential requirements are adequate modelling of clinical pathways as well as usage of adequate systems, which are capable of assisting the complete path of a patient within a hospital, and preferably also outside of it, in a digital way. To fulfil these specifications the authors present a suitable concept, which meets the challenges of well-structured clinical pathways as well as rather poorly structured diagnostic and therapeutic decisions, by interplay of process-oriented and knowledge-based hospital information systems.

## 1 Development of the Process-Oriented Perspective in German Hospitals to the "Clinical Pathways"

Since the mid 1990s process orientation is steadily gaining importance by popular buzzwords like Business Process (Re-)Engineering in enterprises and business managements, and it is nowadays essential basis for goods and services and service offering with the aim of improving the success affecting variables *time, cost* and *quality* [5]. Further important, and often with process orientation associated objectives, are transparency of business processes as well as quality-oriented requirements or certificates [1].

Contrary to the establishment of process orientation in most industry sectors, it could not gain importance in the field of German hospitals until the 2000s. The modelling, analysis and visualisation of workflows in hospitals in the areas of diagnosis, therapy and care proceeded with the goal of getting a detailed overview of the "hospital enterprise".

The initial goal was oftentimes the process acquisition, for example in accordance with the aspired ISO 9001-Certification.[1] During this procedure the business process of the complete enterprise was initially mapped, then in each department the process landscape with the main processes was compiled, and, in a third step, the precise process sequence was surveyed and modelled (cf. figure 1).[2]

---

[1] The obligation to adopt quality management in hospitals originates from §135a, subparagraph 2, sentence 1 of the German Social Security Code Book (*dt. Sozial Gesetzbuch, SGB*). The standards DIN EN ISO 9000ff. are one of the prevalent quality management methods and were developed for the service and production sector [1].

[2] cf. for the relationship between quality- and process-orientation in hospitals also [10], [1].

P. Kostkova (Ed.): eHealth 2009, LNICST 27, pp. 147–154, 2010.

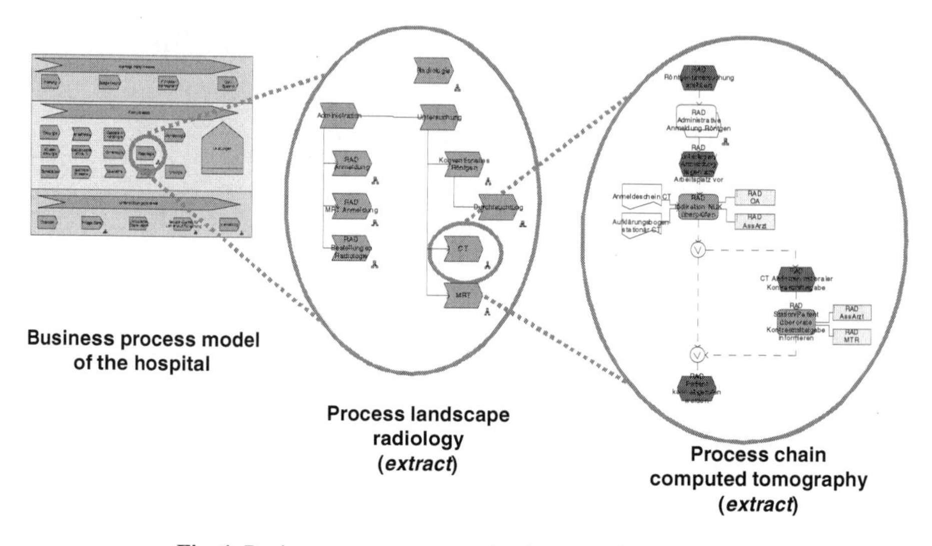


**Fig. 1.** Business process, process landscape and process chain

From an organizational point of view this form of modelling at least succeeded in picturing the detailed layout of the different functional areas (e.g. the clinic for radiology in figure 1), and its service offerings as well as the concrete process sequence of the service offerings, e.g. in the notation of an event-driven process chain (EPC).

Furthermore, the created organization and process model provides the starting basis for process costing as an economic tool for corporate management. To obtain an integrated overall model suitable tools are used. Prerequisite for a successful adoption is the adjustment of the process model to the requirements of process costing, for example by adequate process selection matrices and addition of adequate time and cost parameters. The creation of different process scenarios permitted comparison of process chains from an economic point of view. During process costing the problem-oriented consolidation of available information to key performance indicators and output of result reports takes place, e.g. to spreadsheets or web-reports. Particularly with the adoption of the lump sum compensation settlement or the German Diagnosis Related Group (G-DRG)-System the actual process of creation of value in hospitals takes centre stage. This process is called the *Treatment Path of the Patient, Clinical Path, Treatment Path or Clinical Pathway.*

A Clinical Pathway is an evidence-based treatment flow, crossover professions, considering costs, quality, policy and risk.

The reason for focussing on the clinical path comes from the altered revenue situation. Usually after admittance of patients in hospitals a specific diagnosis is made. Consequentially the assignment to a respective DRG takes place. With it the revenues of this patient are clearly defined.

The expenditures however result from the course of treatment that the patient takes through the different functional areas, and are only very insufficiently recorded. Therefore, to analyse the costs a particular patient or a "case" generates, it is necessary to closely examine the clinical path a patient takes from his admittance to his release.

Respectively, the clinical path acts as a central control for the complete medical course of treatment that ranges over patient care in hospitals as well as integrated concepts (e.g. nursing services or home care concepts). It specifies not only the inherent logic of the process sequence, but also further economic and medical aspects which are explained below.

## 2 Modelling of Clinical Paths

Fundamental prerequisite for the successful modelling and implementation of clinical paths is the use of a modelling syntax and notation that meets the requirements of clinical paths. Therefore, the first step in a business process modelling project consists on the development of the modelling conventions, their specification and the training of involved persons. Furthermore a critical success factor is the selection of an appropriate modelling tool.[3]

During the specification of the modelling conventions the utilised model types, the layer structure of modelling, the objects and their attributes have to be specified. In the specification of the model types focussing especially on the days of treatment is of great importance. Usually the first day is the day of admittance and the last day the day of release, whereas the administrative tasks on both days are less dependent on the respective path. Also the time differences of medical diagnosis are rather small. Oftentimes, on the next day after the day of admittance the operation takes place, which consists of the preparation, the procedure and post-processing of the operation. The other days are lay days, which are under minor variation. In accordance to this structuring the design pattern of a process chain with a column layout with rough presentation of the activities is suitable for the super ordinate level (main process sequence of the clinical path). Additionally the columns are structured according to the days of stay or to the activities that are performed on those days. The first day is the admittance of the patient; the treatments are on days 2 to 8, followed by the release. The activities on days 2 to 8 differ in whether the patient is to be treated in standard or intensive care unit. Accordingly, the function "standard care unit" contains the process chain, which is processed each day of treatment on each patient. Structurally, it is located on the second hierarchical modelling level. The patient receives different therapies on each day, like inhalation or breathing exercises. The detailed process description, which contains activities like the procedure of inhalation, is not displayed on this level. Rather the displayed level shows the composition or configuration of the individual "treatment modules" of the clinical path, which are specific to the path. The activities that have to be performed during the treatment are displayed as an event-driven process chain on the third and thus most detailed level. The following figure 2 visualises the described modelling example.

The explained structure shows the advantages of the concept. Because of the modular structure the range of services of a clinic or department of the hospital can be uniquely illustrated and is integrated into the respective clinical path. Therefore, during the development of policies that represent a clinical path, the acquisition and illustration of the

[3] For example the ARIS Toolset by IDS Scheer AG is suitable as a professional modelling tool because of different characteristics, like the widespread use, the widespread recognition as a virtual standard, the client-server architecture and the close connection methods.

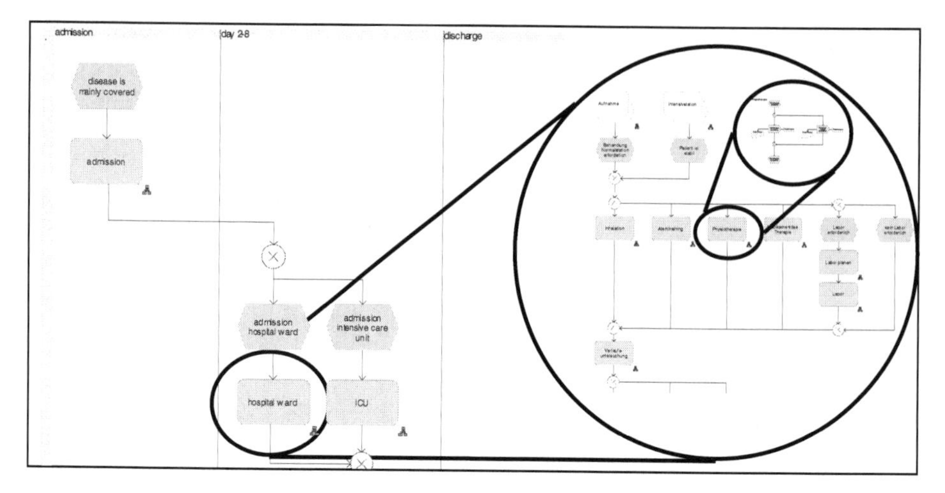


**Fig. 2.** Clinical Path Pneumonia

respective treatment modules come first. These modules in turn are referenced or integrated into the respective paths. With this the model of the clinical path meets the requirement of being a reticular and comprehensive course of treatment. Besides the dynamic illustration of the treatment paths, naturally also the static illustration of the organizational structure of the hospital and the departments and clinics are made.

With the goal of creating a homogeneous and consistent overall model, it is necessary to define the number of necessary objects for modelling appropriately. Therefore, the stocktaking of the resources has to be carried out initially to determine appropriate objects based on functional and technical modelling criteria. The following Table 1 provides an exemplary overview.

**Table 1.** Overview of Modelling Objects (selection)

| Tag symbol | Object type | Usage |
|---|---|---|
| event | Event | Triggers function or is result of triggered function |
| function | Function | Activity that processes data or material and needs time or ressources for that |
| performance index | Performance figure | Is used in the attribute "Description/Definition" to indicate descriptions and notes on expenses |
| risk | Risk | Describes the risk of the activity including risk assessment and control measures |

In the presented selection of objects the total number of objects provided for modelling was kept as small as possible, with the aim of creating well structured and comprehensible models. Especially in very extensive projects with involvement of employees from the functional areas this approach is suitable. Besides the visualisation of the path traversal in the form of a process model, the extension of the objects of the model with additional attributes is necessary. The attributes can for example contain additional descriptive information for the future users of the model (e.g. physicians or nurses). In addition, special attributes can also be used for further analysis or simulations, such as process costing. Respectively, while specifying the modelling conventions an overview of each object has to be created that shows, which attributes (optional or mandatory) have to be maintained and with which values or contents these have to be filled.

When using powerful and extensive modelling tools such as the ARIS Toolset, it is sensible to create suitable and problem-oriented model templates based on the method-handbook to assist the consistent implementation of the modelling conventions. Furthermore, it should not be possible for the modeler in the department to create new objects. Rather every possible object is to be collected in an as-is-analysis and created one-time to work with the instance of this object during modelling. This approach additionally aids the consistency of the models and prevents rank growth or also redundancy of objects.

# 3 Architectural Concept of a Knowledge-Base Process-Oriented Hospital Information System

Besides the modelling and adoption or implementation of clinical paths the support by information systems of the path traversal presents a major challenge for the clinic. In fact existing hospital information systems support only particular function-oriented stations of the complete path and offer no support for the overall process. When using a process-oriented HIS the holistic support of the clinical path of the patient and hence all diagnosis, therapy and nursing processes of a patient take centre stage and hence the necessity for information-technological support of all tasks and activities during the mentioned clinical processes.

The challenge of designing, implementing and using of process-oriented HIS is the detailed depiction of processes as well as the integration of information-oriented and functional application and information systems in the hospital. In addition to the usage of the existing path and process models in hospitals as well as their itemization and customization the integration of existing applications is necessary for the technological implementation. Besides the process-oriented electronic support of the clinical treatment path and the integration of data- and information-oriented applications other challenges exist in the configuration of the control layer as well as the creation of a suitable user- and role-model.

## 3.1 Limits of Process-Oriented Hospital Information Systems

An extension of the described architectural concept is necessary to allow active support for the processes. From the process-oriented perspective the focus is primarily on the flow of the system. While the process chains described by the clinical path are well structured and suitable to be supported by a workflow management system, problems arise on the interfaces between the clinical paths (for instance entering a path or change of path) because of the rather poor structuring. A multitude of information (e.g. laboratory findings and vital signs) coincides at this point, which are of high importance for further treatment. Because of the multitude of alternatives, rules and decisions to be processed a workflow system is rather unsuitable in this case. In fact it is sensible to use a system because of its strength, which possesses its own knowledge, can process incoming information and conclude new decisions. The implementation of such a knowledge-based system onto the described interfaces would contribute further digital support to the treatment path of the patient. For instance, after admittance of a patient different examinations (e.g. radiological diagnostics and examining/measuring of vital- and blood values) are performed. These examinations constitute amongst others the starting point for diagnosis making. After performing the examinations the (partly) automated evaluation of the results or the assistance of the physician in evaluating the results takes place. After the evaluation of the partial results the diagnosis is completed, so that the therapy process can begin. With the described knowledge (for example in form of decision-making rules) one or more therapy processes or also the use of other diagnostic procedures are recommended, at which the level of information is supplemented with the rating of the alternatives and their respective probabilities and explanatory notes. These pieces of information serve or support the physician in his task to chose a suitable clinical process for the further treatment of the patient. Besides the usage and the additional support for the clinical process it should be possible for the decision maker to modify the existing decision-making rules or their weight, and therewith to contribute to the improvement of the system and expand its knowledge base.

### 3.2 Architectural Concept of a Knowledge-Based Hospital Workflow-System

From the described requirements for the architecture of a process- and decision-oriented HIS follow the necessity for the integration of a suitable system component that has the ability to use knowledge (e.g. in form of rules) on incoming data or information, come to a conclusion and describe the path to that conclusion as well as extend the existing knowledge based on the processed information. To meet these requirements the use of *knowledge-based systems (KBS)* is especially adequate. Its usage in the field of medicine so far has been especially successful in diagnostics in different fields of applications [6], [7]. For the implementation of the herein described requirements a much more open and universal system is necessary, which is equipped with the necessary components in terms of a common and ideal-typical architecture of a KBS [2].

The result of combining a knowledge based system with a process oriented system is an active knowledge- and process-oriented healthcare information system, visualized in the following figure 3.

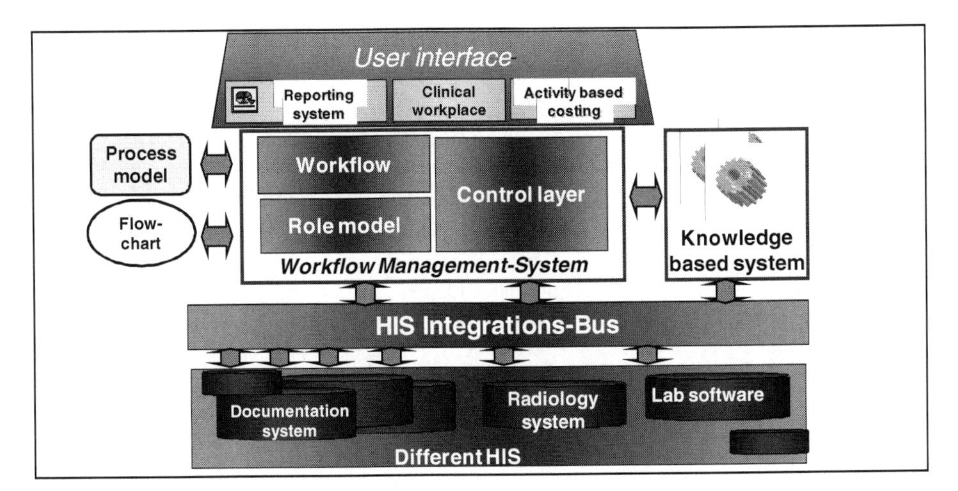


**Fig. 3.** Design model of an active knowledge- and process-oriented healthcare information system

From the illustrated architecture and alignment of the components follows the dominating role of the workflow management system for the complete concept. For instance, from the user's point of view the physician accesses the process-oriented system and receives the presentation of the results. The knowledge-based system provides beyond that no additional user environment; it rather takes over the role of the subsystem within the complete architecture. The same applies for the access to the knowledge-based system in the process sequence: the process-oriented system is the trigger and incorporates the KBS into the process within the process sequence. Thus, the KBS acts as a supporting subsystem that is triggered by the process-oriented system, then it incorporates further information and finally returns a result to the process-oriented system.

## 4 Synopsis and Future Prospects

The competition on the German hospital market is in full swing, and the progress on the health care market in Germany will lead to only efficiently and economically working enterprises prevailing.

Contrary to the predominant technology penetration in hospitals the idea of process orientation and especially the adoption of process-oriented hospital information systems have not yet been realized. For the coming years support by process-oriented and knowledge-based information- and communications technology will prove to be a critical factor for success.

On the other hand suitable approaches of the market leaders in this segment are still under development. With the described concept the authors provide a substantiated approach in practice to overcome the problems associated with process orientation in hospitals, and to present an efficient and suitable architecture as a solution for the problem.

## References

1. Bohr, N.: Prozessmanagement als Grundlage für Zertifizierungen nach EFQM und ISO. In: Braun, G., Güssow, J., Ott, R. (Hrsg.) (eds.) Prozessorientiertes Krankenhaus, Stuttgart, pp. 181–194 (2005)
2. Gabriel, R.: Wissensbasierte Systeme in der betrieblichen PraxiS. McGraw-Hill, London (1992)
3. Gabriel, R., Lux, T.: Decision Support Systeme im Krankenhaus – Aufbau eines wissensbasierten und prozessorientierten Krankenhausinformationssystems. In: Bortfeld, A., Homberger, J., Kopfer, H., Pankratz, G., Strangmeier, R. (Hrsg) (eds.) Intelligent Decision Support: Current Challenges and Approaches, Wiesbaden 2008, pp. 337–357 (2008)
4. Hellmann, W.: Klinische Pfade: Konzepte, Umsetzungen und Erfahrungen (2002)
5. Hammer, M., Champy, J.: Reengineering Work: Don't Automate, Obliterate. Harvard Business Review 7/8, 104–112 (1990)
6. Lux, T., Schneppat, M.: Prozessorientierte und Wissensbasierte Systeme im Krankenhaus, Bochum (2007)
7. Park, Y.-J., Kim, B.-C., Chun, S.-H.: New knowledge extraction technique using probability for case-based reasoning: application to medical diagnosis. Expert Systems 23(1), 2–20 (2006)
8. Prokosch, H.U.: KAS, KIS, EKA, EPA, EGA, E-Health: Ein Plädoyer gegen die babylonische Begriffsverwirrung in der Medizinischen Informatik. Informatik, Biometrie und Epidemiologie in Medizin und Biologie 32, S371–S382 (2001)
9. Sunyaev, A., Leimeister, J.M., Schweiger, A., Krcmar, H.: Integrationsarchitekturen für das Krankenhaus – Status quo und Zukunftsperspektiven. Information Management & Consulting 21(1), 28–35 (2006)
10. Zaugg, B.: Nutzen des Prozessmanagements für die Einführung eines Qualitätsmanagements. In: Braun, G., Güssow, J., Ott, R. (Hrsg.) (eds.) Prozessorientiertes Krankenhaus, Stuttgart, pp. 129–144 (2005)

# With Intègre®, Leverage Every Medical Professionals' Skills and Expertise

Denis Pierre

NORMIND,
cap omega, rond-point Benjamin Franklin,
CS 39521, 34960 Montpellier cedex 2, France
pierre@normind.com

**Abstract.** Intègre® is a decision-support software specially designed for collective processes. Collecting all required information for decision making, Intègre harmonizes representations and processes collected data to check users' decisions coherency. During their consultation, specialist doctors access to analysis' results and to patients' data, stored in their own databases or in administration's systems. Then they get supported by Intègre to produce diagnosis in total coherency with medical guidelines and existing information.

**Keywords:** computer-aided decision making, electronic guidelines, evaluation.

Main difficulties of practice's guidelines permanent use by professionals are guidelines integration into real practice and their update. Moreover, updates, whatever their frequency, have to be adapted with practices and treatments, for a perfect adequacy between referential and practice.

Solving these 2 issues requires to design a referential model which can accept some« gaps » (or norms' violations committed by professionals), which can detect them and exploit them in order to update guidelines and impulse practice evolution.

That's why referential representative existing models can't work :

- « alerts based models » use logic forms which do not accept any contradiction, or any rules violation,
- decision trees models propose an a priori method which doesn't deliver a valid answer in most cases and which update is complex (see previous paragraphs).

As a matter of fact, reasoning suppose to :

- be able to evolve in a logical structure without having all the elements in hand (unknown data)
- accept incoherencies and therefore to search for borderline solutions
- propose multiple choices, to explore other/new options
- allow to evolve freely in the decision tree, without imposed entrance or exit points
- be able to imagine new logic associations and new possible exits.

Normind proposes another direction using logical forms which accept contradiction and keep on assisting decision-making process, while professionals ignore some referential rules. More than a practice guide, guidelines get used to design assumptions

P. Kostkova (Ed.): eHealth 2009, LNICST 27, pp. 155–156, 2010.

and to select the best decision. Guidelines come to be a real help for deciding, and are no longer constraints to be respected.

Thereby expert systems only play their role when it is about pure decisional logic whereas Intègre technology by Normind assists and guides experts reasonings.

Professional medical people's issue is to set up a virtuous circle meeting the numerous standards and requirements of quality, scientific and administrative programs. Such a process must mix practice policies and data management in order to reach goals as follow :

- To support daily practice, using guidelines and categorizing pathologies
- To manage data collection with standards and expected minimal data
- To run professional practices' assessment
- To detect dysfunctions
- To correct practices and guidelines the right way.

Aiming to take an adequate therapeutic decision, Intègre operates data from different files (internal, administrative…) and supports consultation notes' edition, using best practice guidelines.

Confronting real practices to concerned guidelines in a given context (specialty, pathology) allows to manage the patient's case the best way, orienting it in the most relevant and appropriate commissions.

Based on consultation notes and commissions' choices, Intègre helps to analyze practices through an assessment tool.

- Professional practices assessment based on collected data quality, real practice and referential confrontation, minimal required data sufficiency or practices compliance with recommendations
- Clinical best practice update through impact analysis, updating (new facts) and weak points detection (proposal never or rarely applied, rules never respected, proposals for variables correction …)

Intègre offers additional functions to professional tools. Doctors shall keep their existing tools for patients' files management and diagnosis support ; Intègre adds technical elements to link these tools. At any stage, the involved persons fully control the decision making process.

Intègre is a commercial product that will be fully demonstrated during the conference. The decision-support process will be demonstrated with the implemented french prostate cancer guideline, embedded in the software.

# Personality Diagnosis for Personalized eHealth Services

Fabio Cortellese[1], Marco Nalin[2], Angelica Morandi[2], Alberto Sanna[2], and Floriana Grasso[1]

[1] University of Liverpool, Department of Computer Science
Ashton Street, Liverpool L69 3BX, UK
fabiocortellese@gmail.com, floriana@liverpool.ac.uk
[2] Fondazione Centro San Raffaele del Monte Tabor, eServices for Life and Health
Via Olgettina 60, 20132 Milano Italy
{marco.nalin,angelica.morandi,alberto.sanna}@hsr.it

**Abstract.** In this paper we present two different approaches to personality diagnosis, for the provision of innovative personalized services, as used in a case study where diabetic patients were supported in the improvement of physical activity in their daily life. The first approach presented relies on a *static clustering* of the population, with a specific motivation strategy designed for each cluster. The second approach relies on a *dynamic population* clustering, making use of recommendation systems and algorithms, like Collaborative Filtering. We discuss pro and cons of each approach and a possible combination of the two, as the most promising solution for this and other personalization services in eHealth.

**Keywords:** Personalization, Personality Diagnosis, Motivation Strategy, Collaborative Filtering, Natural Language Processing, Contextualization, Dynamic Clustering.

## Introduction

According to the World Health Organization, modifiable behaviours, including specific aspects of diet, overweight, inactivity, and smoking, account for over 70% of stroke and colon cancer, over 80% of coronary heart disease, and over 90% of adult onset diabetes [1]. Intervention trials recently showed that a correct diet in combination with exercise programs can reduce the risk of developing diabetes by 60% in subjects with impaired glucose tolerance [2]. Also, great emphasis has recently been put on improving quality of life for diabetic patients, especially in terms of physical activity [3]. Most people are aware of healthy recommendations (e.g., about diet, physical activity), but they often find them very difficult to put into practice in their daily life, or fail to associate daily micro-behaviours (e.g., driving to work, rather than cycling or walking) with long-term health consequences. It is therefore important to understand how to promote a behaviour, like physical activity, taking into account the psychological determinants of such behaviour.

The European research project PIPS (Personalised Information Platform for Health and Life Services) [4] investigated the use of a eHealth platform for health promotion. In an intervention aimed at promoting physical activities among diabetic patients, a

P. Kostkova (Ed.): eHealth 2009, LNICST 27, pp. 157–164, 2010.

platform was designed, together with a methodology, which used a feedback based support system for the improvement of personal performances in physical activity [9]. This methodology relies on a pedometer for correctly assessing the patient's daily activity, and on a motivational strategy to provide a personalized support. The designed strategy is specific for a particular cluster of population (that have in common a specific *health personality* or, in the specific case, a common *motivational status*), which was *statically* determined by San Raffaele Hospital psychologists. During the PIPS project, some attempts to introduce a *dynamic* clustering have been proposed with promising, yet partial, results. This paper will present first the static clustering approach, describing the Motivational Strategy implemented in PIPS, then it presents a possible improvement using Recommendation Systems, and how it these could be implemented in PIPS. Finally the paper discusses some possible combination and comparison of the two approaches.

## Static Clustering: Motivation Strategy

The motivational strategy designed in the context of PIPS was personalized along several dimensions and delivered through ICT solutions and devices, allowing to constantly support the patients during their daily lives; the proposed strategy has been designed with the aim of exploring the effectiveness of an e-health platform jointly with appropriate motivational tools for health promotion.

The level of personalization in the support tools has been implemented including socio-demographic and individual characteristics. PIPS introduces a motivation assessment determining the stage of behaviour change the user is at, as well as a detailed profiling and medical assessment. PIPS then provides a personalized and incremental target in terms of number steps, walking time, speed and caloric consumption and gives the patient a pedometer, that has been demonstrated to be a valid monitoring and motivation tool. With the pedometer, walking data are constantly updated and corrective, motivational messages are delivered just-in-time to the user mobile phone, thus supporting patient compliance with a real time response.

Motivational Messages are sent, with the aim to give feedback on the performance and to motivate to improve, giving advice to support their walking activity.

During the course of the program the patients receive several kinds of motivational message: a standard message is sent Monday to Thursday; a Friday *Special* message includes suggestions for the weekend; a Sunday *Summary* provides an overall judgement on the week performance. Moreover, a *Recover* message is sent when the patients significantly underachieve the daily target, around the time when they are supposed to have completed the activity they committed to: this message is an alert but also contains advice on how to achieve the target before the end of the day.

Messages are personalized along several dimensions: motivational stage of change, performance level, emotional status, how far is the patient in the programme, and other user's features as indicated in the profile (e.g. dog owner, drive to work, etc), location (e.g. weather forecast), perceived obstacles, as indicated by the patients when they fail to achieve their targets.

On the last day of the month the patient also receives a report of the month, which is intended to provoke some thoughts on the reasons of success/failure, so as to increase self awareness and eventually modify personal strategies, and the patient can comment on the system's deductions. Finally, the system sends a message containing a proverb related to the positive/negative factors that influenced the target achievements or a suggestion/encouragement for the next month.

A message structure was identified consisting of segments, each of which related to personalization factors and/or communicative goals [10]. The generation of the messages is achieved by means of a composition algorithm using constraint satisfaction techniques, selecting among around 1000 canned text segments, which are filled in with data from the database.

The PIPS Walking Program underwent different validation phases throughout the whole implementation process. In a first, informal evaluation, diabetes medical doctors and psychologists were asked to comment on the pilot. A second evaluation involved about 50 patients selected within the Outpatients Diabetes Care Centre of San Raffaele Hospital aged between 45 and 70. Patients had the opportunity to wear the pedometer for 15 minutes, to see the graphical representation of their walking performance and to see examples of the personalized messages composed after entering some selected information (e.g. diary, preferences, habits). Users were asked to feedback about the presented services: e.g. usability, information effectiveness, messages motivational level and understanding, exercise plan usefulness, etc. Of the patients interviewed, 75% said they were more inclined to increase their physical activity and to use technological devices as a support to the diabetes management.

Currently, a National, mono-centric (Diabetes, Endocrinology and Metabolism Department, FCSR, Milan), randomized, open-label, intervention study is ongoing enrolling patients from San Raffaele Diabetes Outpatient department. The study protocol, approved by San Raffaele Ethical Committee, sees the enrolment of 60 patients for a duration of 6 months each (2 run-in weeks, 3 months intervention period, 3 months control period), randomized according to two branches: the control group receives PIPS Pedometer as monitoring tool and standard diet/exercise care, the intervention group receives PIPS pedometer and mobile phone, a personalized walking target path (steps/min and total minutes) and information and motivation feedback. Inclusion criteria consider patients with Diabetes Type 2, aged 35-70, with no physical or psychological impairment and with a BMI<35. The primary objective of the study is to demonstrate the effectiveness of PIPS, which integrates a technological platform and a personalized motivation strategy, to achieve a personalized exercise target and to improve patient compliance. Patients were profiled according to the stage of behaviour change both towards exercise and towards technology [11]. First results report an improvement in the metabolic profile of the patients after 3 months of physical activity supported by PIPS strategy. Feedback from users indicates that the tool overall increases their willingness to walk. With respect to the messages, they are considered relevant to the actual target achievement, but sometimes too repetitive, so they were read with less interest over time.

## Towards A More Effective Solution: Dynamic Clustering

One of the shortcomings of the aforementioned solution is that the choice of a message among a set of equally relevant ones does not take into account the projected outcome of this message in terms of improving the patients' performances. While an accurate measure of the message effectiveness would require a very sophisticated user model, including also emotional and cognitive factors, some prediction could be achieved by using knowledge coming from other users, or the same user in other similar situations. Our proposal is therefore to apply to the choice of which message to present the user with, techniques coming from works on recommender systems. In particular, our assumption relies on a Personality Diagnosis [12], that is the probability that a user is of the same "personality type" as other users, influences the probability that the user will adopt a behaviour (e.g. like an item, buy a product, etc).

The typical recommender system has the aim of predict user's preferences towards some products to buy or examine, on the basis of information acquired on the community of the system's users. Algorithms in use can be of various sorts [5][7][8], and come from diverse areas, like information retrieval, information filtering, data mining, or machine learning. Recommender systems can use collaborative filtering (based on the relationships among users), association rules (based on the relationships among products) or classifiers (based on the content of the knowledge base).

For example, in a system recommending books there usually are two sets, a set of users $U$ (e.g. the readers) and a set of items $I$ (e.g. the books), and a utility function $r$ between the two sets (e.g. the rating the users gives to the books). In its most common formulation, the recommendation problem is reduced to the problem of estimating ratings for the items that have not been seen by a user, selecting for each user $u \in U$ the item $i' \in I$ that maximizes the defined user's utility.

In our specific case the set $U$ of users is the set of diabetic patients, while the set of items $I$ is formed by the set of all the possible messages to be sent to the patients. The relationship between the two sets, instead of being a rating, is a "success" function. The hypothesis is that, if a motivational message was "successful" we would expect performances to improve. The success function is therefore determined by the walking results, measured as the percentage of achievement of the target on the following day of walk.

Messages, as well as the patient profile, can be considered as vectors, described by a set of parameters. For example for the message we can consider the parameters:

- type (messages can give a strong o a mild encouragement),
- subject (some of them are related to the body, other to health in general, other to social aspect of physical activity, etc.),
- length (different person may like short messages, other complete information),
- value (some messages are more positive, e.g. "if you walk you will feel the benefit", other are more negative, e.g. "if you don't do you physical activity you will have complications…", etc),
- attitude (some message can be more friendly, other more formal).

In this case we can present each message $i \in I$ as:

$$i = (i_1, i_2, \dots, i_m)$$

For the patient, we can consider the dimensions explained in the previous section, for example motivational stage, performance level, preferences, perceived obstacles, etc. In this way we can consider each element $u \in U$ as a vector described by its dimensions:

$$u = (u_1, u_2, \dots, u_n)$$

With this model, two recommendation services are possible:

1. the patient will be provided with messages similar to the ones that worked better in the past (content-based recommendations);
2. the patient will be provided with messages that worked with people with similar characteristics and preferences in the past (collaborative recommendations).

In content-based recommendations systems the goal is to identify similarities between items, and that's the final purpose. In collaborative recommendation this is just an intermediate step used to identify similarities of "tastes" between users. In our case study, the first service model seems preferable, because it has the advantage that the cold start problem, typical of the recommendation systems, is limited: the patient receives at least one message per day (more if underperforming), a Friday special, a Sunday special, and the monthly report. In average we can consider from 40 to 70 messages per month, or 480 to 840 messages per year, which makes a lot of data available to the system.

To calculate similarity, a recommendation systems typically takes into account all items that are co-rated by two different users. In our case we will choose the subset of $I$, $I_{xy} = I_x \cap I_y$, where $I_x$ are all the items rated by patient $x$ and $I_y$ are all the items rated by user $y$. Collaborative Filtering algorithm can determine the nearest neighbours of user $x$ without computing $I_{xy}$ for all users $y$. In the correlation-based approach, the Pearson correlation is used to calculate the similarity [7][8]:

$$sim_{Pearson}(x, y) = \frac{\sum_{i \in I_{xy}} (r_{x,i} - \bar{r_x})(r_{y,i} - \bar{r_y})}{\sqrt{\sum_{i \in I_{xy}} (r_{x,i} - \bar{r_x})^2}}$$

In the cosine-based approach, the two users $x$ and $y$ are treated as two vectors in m-dimensional space, where $m = |I_{xy}|$. Then, the similarity between two vectors can be measured by computing the cosine of the angle between them:

$$sim_{cos}(x, y) = \cos(\vec{x}, \vec{y}) = \frac{\sum_{i \in I_{xy}} r_{x,i} r_{y,i}}{\sqrt{\sum_{i \in I_{xy}} r_{x,i}^2} \sqrt{\sum_{i \in I_{xy}} r_{y,i}^2}}$$

The best similarity function has not been identified yet, because of the lack of sufficient data in the study to compare the two approaches.

## Implementation Of The Personality Diagnosis Method In PIPS

The basic idea is, as said above, to consider the appropriate message to present to a patient as the most "recommendable" message, on the basis of how the message to propose (or others in a similar class) was evaluated in the past either by the same patient or by other patients in the same situation of the user. A good evaluation is considered, broadly speaking, one when, after receiving the message, a user did improve his performances: so when the target patient has similar characteristics and similar past performances, one can assume the same message may provide the same outcome. The interaction of a user with a typical recommender system is however explicit: the user browses some products, or purchases them, or provides an evaluation for them. In our case, the only way in which a patient can interact with the system is by providing the daily performance, or updating his data. The approach used, therefore, is the one of creating a relationship among the performance and its derivative, the messages, and other data collected about on the users.

In PIPS, the characteristics of the message to recommend come from an ontology [10], so relevant classes in the ontology are used as parameters. The message itself is of course not the only factor that influences the performance of the patient: the context like the weather for the day or how good was the day at work, can be determinant of the performance no matter how good the motivational message was. It made therefore sense to consider, in computing the evaluation, both the *message* and the *context*. Thus, $I' = I \times C$ is the new extended item set for our recommendations, where $I$ is the set of all the possible messages and $C$ is the set of all the possible contexts. Formally we can say that each element $i' \in I'$ has the following components:

$$i' = (i_1, i_2, \dots, i_m, c_1, c_2, \dots, c_p)$$

The idea of extending the recommendation by including contextual information was proposed also for a movie recommendation system [5], where the suggestions showed meaningful improvement by adding contextual information, such as when, where, and with whom a movie is seen. In our preliminary implementation we considered as context the parameters that the patient inputs on a daily "diary". The patients can input aspects related to their day as emoticons, choosing to comment on the level of gratification at work on that day, their perception on the social relationships on the day, the weather, and an overall emotional orientation, or "mood". This is a simple solution, but one can think, assuming to be able to rely on monitoring system and on internet public services, of adding more descriptive contextual variables, such as environment (e.g. pollens concentration), physical parameters like blood pressure; social interactions, from events in the user's day, etc.

In the current implementation, the system considers the performance of the day, in relation to the one on the previous day. If the performance improves, the message of the day before is "promoted", otherwise it is "demoted", of a factor depending on the context, given by entries in the user's diary. Entries in the diary with a positive orientation (e.g. a good day at work) cause a lesser increase in the value of the message when the performance improves (the idea being that the user might have been motivated by the positive experience, rather than the message), and cause a greater decrease when the performance deteriorates (representing the fact that the message was not effective in spite of the positive attitude). Similarly, entries in the diary with a negative orientation (e.g. a bad day at work) cause a greater increase in the value of the message when the performance improves, and a lesser decrease in the value of the message if the performance deteriorates.

The work on fine tuning the formulae is still ongoing, and is based on the result of preliminary questionnaires where patients were asked to evaluate the messages produced with the original method. The objective of this analysis, for our purposes, is to obtain some *a priori* values for starting the system up, and possibly a set of benchmark values for evaluation.

## Final Considerations

In this paper we presented two possible approaches for determining the health personality of patients, with the purpose of tailoring specific personalized strategies and services. The case study presented, in particular, describes a service for physical activity support, where (diabetic) patient receives motivational feedbacks to improve their performances. Static and dynamic clustering are important bases for providing personalized services. Both approaches present advantages and disadvantages.

The Static clustering has the advantage that the reference model (e.g. the motivational status psychological theory) is known a priori. Correlation between relevant factors is identified from the start and the support feedback is designed to maximise effectiveness. On the other hand, if the model is not descriptive enough of the process, or if many uncontrolled variables influences the final results, the static approach is less personalised and contextualised, and may become repetitive.

In dynamic clustering the correlations between relevant variables are not known a priori, but, on the other hand, they become clear during the processing and can change when new data is taken into account. The ability to cluster the population based on patients' behaviour makes this approach very promising to predict future behaviours, thus to provide the proper support feedback. The dynamic clustering suffers from all the problems typical of recommendation systems, in particular from the *cold start*, when there are not enough data about the users or about messages.

PIPS started the clinical trial described above making large use of a Static Clustering approach, that classified the patients' health personality based on their *motivational status* and provided specific messages, validated by psychologists, to bring a change in the patient's motivation and actions. A first attempt was also made to use also the dynamic clustering and, even if results are still very preliminary, they seems very promising. The lesson learned is that probably the best solution would be a combination of the two clustering. This, in our case study, can be done in two ways:

1. Using the static clustering when a new user enters the system, until there are sufficient data about him to start using the dynamic clustering. This approach has the advantage to reduce the problem of the cold start of the recommendation system, and that in the second phase it will fully adapt to the behaviour of the user. On the other hand, messages provided to the user in the second phase are *uncontrolled* by medical personnel, thus a patient may receive a message that would not have been selected for the same patient by the static, controlled method, so a message which could not have been pre-approved by the medical experts.
2. Using the dynamic clustering on a set of messages that have been preselected to be appropriate to the static cluster to which the user belongs. In other words, we apply first the static clustering, preselecting the subset $I_x \subset I$, of all the items (messages) appropriate for user $x$ based on the users' motivational status. We can then use this subset of messages to calculate the new item set $I'_x = I_x \times C$, as the item set to which apply the collaborative (finding similar users) or content based (finding similar messages) recommendations. This approach is safer from the point of view of medical validation, but will lead to less variability.

In both cases, it is crucial to ensure that the monitored variables are descriptive enough of the factors impacting on user's decisions and behaviours, and are as complete as possible.

The two approaches were compared in terms of complexity and effectiveness with respect to the case study. A formal evaluation of the first approach is currently under way with a randomized trial, while for the second it is envisaged a less formal evaluation, due to the experimental nature of the approach.

## References

1. World Health Organization (WHO). Library Cataloguing-in-Publication Data. The global burden of disease: 2004 update, ISBN 9789241563710
2. Swartz, A.M., Strath, S.J., Bassett, D.R., Moore, J.B., Redwine, B.A., Groer, M., Thompson, D.L.: Increasing daily walking improves glucose tolerance in overweight women. Preventive Medicine (37), 356–362 (2003)
3. Delahanty, L.M., Conroy, M.B., Nathan, D.M.: Psychological predictors of physical activity in diabetes prevention program. Journal of American Dietetic Association (106), 698–705 (2006)
4. PIPS - Personalized Information Platform for Life and Health Services. Contract N°: IST-507816, http://www.pips.eu.org
5. Adomavicius, G., Sankaranarayanan, R., Sen, S., Tuzhilin, A.: Incorporating Contextual Information in Recommender Systems Using a Multidimensional Approach. ACM Trans. Information Systems 23(1) (January 2005)
6. Hill, W., Stead, L., Rosenstein, M., Furnas, G.: Recommending and Evaluating Choices in a Virtual Community of Use. In: Conf. Human Factors in Computing Systems (1995)
7. Resnick, P., Iakovou, N., Sushak, M., Bergstrom, P., Riedl, J.: GroupLens: An Open Architecture for Collaborative Filtering of Netnews. In: Proc. 1994 Computer Supported Cooperative Work Conf. (1994)
8. Shardanand, U., Maes, P.: Social Information Filtering: Algorithms for Automating 'Word of Mouth'. In: Proc. Conf. Human Factors in Computing Systems (1995)
9. Morandi, A., Serafin, R.: A personalized motivation strategy for physical activity promotion in diabetic subkects. In: Cawsey, A., Grasso, F., Paris, C., Quaglini, S., Wilkinson, R. (eds.) 2nd Workshop on Personalisation for eHealth, User Modelling Conference (2007)
10. Erriquez, E., Grasso, F.: Generation of Personalised Advisory Messages: an Ontology Based Approach. In: 21th IEEE International Symposium on Computer-Based Medical Systems, Jyvaskyla, Finland, June 19-21, pp. 437–442. IEEE Press, Los Alamitos (2008)
11. del Hoyo-Barbolla, E., Kukafka, R., Arredondo, M., Ortega, M.: A new perspective in the promotion of e-health. Stud. Health Technol. Inform. 124, 404–412 (2006)
12. Pennock, D., Horvitz, E., Lawrence, S., Lee Giles, C.: Collaborative Filtering by Personality Diagnosis: A Hybrid Memory- and Model-Based Approach. In: Proceedings of the 16th Conference on Uncertainty in AI (UAI 2000), pp. 473–480. Morgan Kaufmann, San Francisco (2000)

# Collaboration through ICT between Healthcare Professionals: The Social Requirements of Health 2.0 Applications

Pieter Duysburgh and An Jacobs

IBBT – SMIT / VUB, Pleinlaan 9, 1050 Etterbeek, Belgium[1]
Pieter.Duysburgh@vub.ac.be, An.Jacobs@vub.ac.be

**Abstract.** Social requirements are defined as the users' needs related to the use of an application in interaction with others. This paper aims to formulate social requirements of health 2.0 applications for professional healthcare workers. Collaboration is seen as the central characteristic of these applications. To detect the social requirements, we first identified four features that determine how healthcare professionals collaborate: (1) the professional status of healthcare professionals; (2) patient centeredness; (3) ambiguity in medicine and (4) complex organisation of healthcare. Based on these characteristics and findings of Computer-Supported Cooperative Work (CSCW) research in healthcare, we were able to formulate three social requirements for health 2.0 applications: (1) supported autonomy; (2) rationale in context; and (3) fluid collaboration. These requirements will serve as input for health 2.0 scenarios.

**Keywords:** CSCW, healthcare, social requirements, health 2.0.

## 1 Introduction

### 1.1 Health 2.0

In the aftermath of the dot-com bubble, Dale Doherty argued that all the companies that had survived the collapse had some characteristics in common. These characteristics could be called 'web 2.0' [1]. In no time, it was examined what sort of value web 2.0 applications could add to the domain of healthcare [2, 3].

Currently, definitions of health 2.0 vary widely [4]. From a healthcare point of view, several characteristics that web 2.0 enables and facilitates have been identified as especially relevant by eHealth specialist Gunther Eysenbach [5]. He mentions social networking, participation, apomediation (networked collaborative filtering processes), collaboration and openness as characteristics of health 2.0.[2] The health 2.0

---

[1] This research was made possible within the framework of the IBBT (Institute for Broadband Technology) project Share4Health, a collaboration between Flemish university research groups of IBBT and industrial partners (http://www.ibbt.be/nl/project/share4health).

[2] Eysenbach prefers the term 'medicine 2.0': some authors see it as a broader concept than health 2.0 since it more explicitly includes 'science 2.0'. But he quickly adds that most authors do not see a significant difference between the two terms.

P. Kostkova (Ed.): eHealth 2009, LNICST 27, pp. 165–172, 2010.

discourse is diverse and many different interpretations are attributed to the concept [6]. Some see health 2.0 applications as empowering tools for patients: they may change the dynamic between patient and physician and in doing so ideally make healthcare more patient-centred. Others, however, prefer to see health 2.0 applications as a tool to create a more efficient, lower-cost healthcare, a tool for healthcare reform [6]. This research was done as part of the Share4Health project, which aims to develop a collaboration space for healthcare professionals: a platform that facilitates the exchange of clinical data. However, within the project we also look into forms of collaboration that go beyond the mere exchange of data. Therefore we chose to focus less on the empowering qualities of health 2.0 applications, and more on the possible advanced forms of collaboration health 2.0 offers.

In our view, collaboration is the central characteristic of health 2.0 applications. When we look at the health 2.0 as described by Eysenbach, we see openness, participation and social networking as preconditions or partial concepts of collaboration. To collaborate, there need to be at least two people (social networking) who engage (participation) in some sort of exchange of information (openness). Furthermore, the use of networked collaborative filtering processes can only be the result of such collaboration and therefore we see it as a deduction from collaboration.

With the growth of healthcare specialities, it becomes common and even indispensable for healthcare professionals to work in groups in varied settings [7]. Collaboration will have to grow across organizational, cultural and geographical borders, and health 2.0 applications could play a part in facilitating this collaboration.

'Groupware' or applications that facilitate collaboration through ICT have been the focus of a lot of research [8, 9], as it was frequently observed that, while groupware offered interesting possibilities for individuals and organisations, the groupware was not used or not used correctly. Often, it was seen that failed applications ignored the needs of the users. One way to avoid this and to bring in the user's need in the development process of groupware is the formulation of social requirements.

## 1.2 Social Requirements

Social requirements are the users' needs related to the use of an application in interaction with others. Hereby, the users are regarded as a group of people that pursue a common goal, which is embodied in the application that they use. Social requirements focus on social interaction within the use of the application. Therefore, they go beyond the classic human-computer interaction (HCI) perspective.

As we said, our attention for social requirements should be seen as part of a tradition in design research that underlines the importance of user involvement in the design process [10]. These research traditions all refer to the user to drive, inspire and inform the design development process. The main focus in these traditions is to bridge the gap between the user research and the design process, which has been called the *social-technical gap*. Ackerman defined this gap as *"the great divide between what we know we must support socially and what we can support technically"* [9].

More specifically, our approach can be framed in what has been called the 'third paradigm of HCI': a situated perspective where *"the artefact and the context of the artefact are both defining and subject to interpretations"* [11]. This means that we look at the practices and activities within the world of the users the application is aiming at, look at how their interactions can be supported by the application.

How an application should look, operate and feel is an iterative process, a process of mutual shaping between an application and its users. As a result, the process of requirement formulation is also an iterative process: users shape an application, but the new application again changes the users and their requirements.

## 2 Method

This paper was written in preparation of user research to be conducted in the Share4Health project. We took the overview article of Ackerman on CSCW [9] as a starting point. Based on his findings, we scanned the literature and brought all findings on CSCW research in healthcare and collaboration in healthcare together.

This allowed us to identify some characteristics of collaboration in healthcare. Through combining these characteristics with the CSCW findings we were able to formulate some hypotheses on social requirements of health 2.0 applications.

The future user research will be used to further refine the social requirements and to further reformulate these rather abstract and to a certain extent speculative social requirements to more concrete technical and social recommendations.

## 3 CSCW in Healthcare

As we pointed out before, we see health 2.0 applications as a variation of groupware or CSCW in healthcare. Grudin identified eight challenges for adoption and domestication of groupware that are still relevant for web 2.0 application [8]. These challenges shed a first light on the reasons why applications that support collaboration often fail in practice: (1) Groupware often requires additional work of individuals who do not perceive a (personal) benefit; (2) groupware needs a certain number of users before the application proves to be useful; (3) groupware may bring changes in the social structure and social hierarchy, and touch upon social taboos; (4) groupware may not accommodate the exception handling that is inevitably part of group work; (5) group work features are used less frequently than other features, requiring unobtrusive integration with more heavily used features; (6) analysis and evaluation of groupware is difficult, which makes it difficult to learn from experience; (7) design decision makers lack of good intuition for groupware applications, which results in bad product development; (8) the introduction and implementation of groupware requires more care than other applications.

In his 2000 paper *'The challenge of CSCW'* Ackerman further elaborates on these challenges and provides an overview of the CSCW findings that are most relevant for the social-technical gap [9]. We looked at how these findings were applied in the field of healthcare and were able to distinguish four elements that characterize collaboration in healthcare:[3]

---

[3] Although we speak of 'healthcare' as if it were one single system, it is clear that there are large differences in the tasks and work environments of different types of professional healthcare workers. The focus of the Share4Health project has mainly been pharmacists and GPs and that has also been our main focus in the reviewed articles, although we did not excluded information on collaboration between other types of professional healthcare workers when this seemed relevant. Also, due to the early stage of this work, it was not possible yet to focus on a specific context within healthcare.

- **Professional status or the 'art' of being a healthcare professional.** There is a general sense among healthcare professionals that the profession and the status of the healthcare professional is threatened [14]. New tools to support professionals are therefore often regarded with suspicion: Kaplan noticed that certain doctors fear that their professional status will be undermined by ICT tools [15].
- **Patient centeredness.** For most healthcare professionals, a new ICT tool or system should directly benefit patient care. When a system takes up too much time, a doctor might think that he or she *"could have done a hip replacement instead"* [16]. If a tool does not directly benefit patient care, chances are that the healthcare professionals will ignore it, even though it might for example lower healthcare costs, and so help the total population of patients. In other words, benefits incentives are to be considered on every level, both institutional and individual. And while a system might be beneficial for the healthcare system, doctors and pharmacists see patient care as the central task of their job, and will try to avoid most of the administrative tasks.
- **Ambiguity in medicine.** Medicine as a science is characterized by a lot of uncertainty: while most of medicine is evidence based, it is also partly based on interpretation and opinion [16]. Both diagnosis and treatment are often disputable. This makes collaboration more difficult, as more information needs to be exchanged on the reasoning of the healthcare professional and the line of thought that decisions were based on.
- **Complex organisation of healthcare.** Healthcare is increasingly becoming more specialised. This advanced division of tasks results in a higher need for collaboration and information exchange [17, 18].

# 4 Social Requirements of Health 2.0

Based on these characteristics and CSCW findings on collaboration in healthcare, we propose three social requirements for collaborative ICT systems in healthcare, or in this case, health 2.0 applications: (1) 'supported autonomy', (2) 'rationale in context' and (3) 'fluid collaboration'. Before we further elaborate on these social requirements, we wish to underline that these are hypotheses, based on the characterising of collaboration in healthcare and the findings above. Figure 1 illustrates how these hypotheses are related to the characteristics of collaboration in healthcare. These relationships are further explained at the end of each paragraph.

## 4.1 Supported Autonomy

By introducing the social requirements 'supported autonomy' we want to underline the need for support of many healthcare professionals in their profession. The number of self-employed doctors is decreasing because of the lack of support [14]: healthcare is a highly specialised field and its complexity can be discouraging. However, doctors are reluctant to accept support, as they fear that this means that they will also lose control. Doctors fear that extra support comes at a price, and it will take them away from their core task, namely, taking care of the patient [16]. Doctors especially do not see it as their job to take care of administrative tasks.

Doctors highly value their autonomy. This is shown in a study by Dreiseitl and Binder [19] on the use of clinical decision support systems (CDSSs) by doctors. In 24% of the cases the doctor did not agree with the result of the CDSS, especially the more experienced and confident doctors did disagree more easily. From the viewpoint of this paper, we could conclude that doctors value the support they get, certainly when they do not feel certain about a diagnosis, but they feel that they should have the liberty to disagree with the system.

A similar indication of the importance of supported autonomy can be found in a study on the use of a semi-standardised discharge letter as a communication tool between the hospital and the GP [20]. It was noticed that the GP domesticated the discharge letter upon receiving. The GP interpreted the data by highlighting certain parts of the discharge letter. This way, the information became more useful and the GP protected his or her field of expertise [20].

In conclusion, we state that a health 2.0 application should bring a doctor in contact with other healthcare professionals, but at the same time protect his or her individuality. Professional autonomy is highly valued in healthcare and any system that diminishes this autonomy will be regarded with suspicion.

**Relationship to the characteristics of collaboration in healthcare:** As GPs and pharmacists feel that their *role as a professional healthcare worker* is threatened, they want to concentrate more on their advisory role and they need support to fulfil it. When the medical information is *ambiguous*, a GP or pharmacist may in certain cases want to know the opinion of a colleague because he or she is not certain of doing the right thing. This also means that different healthcare professionals can have different

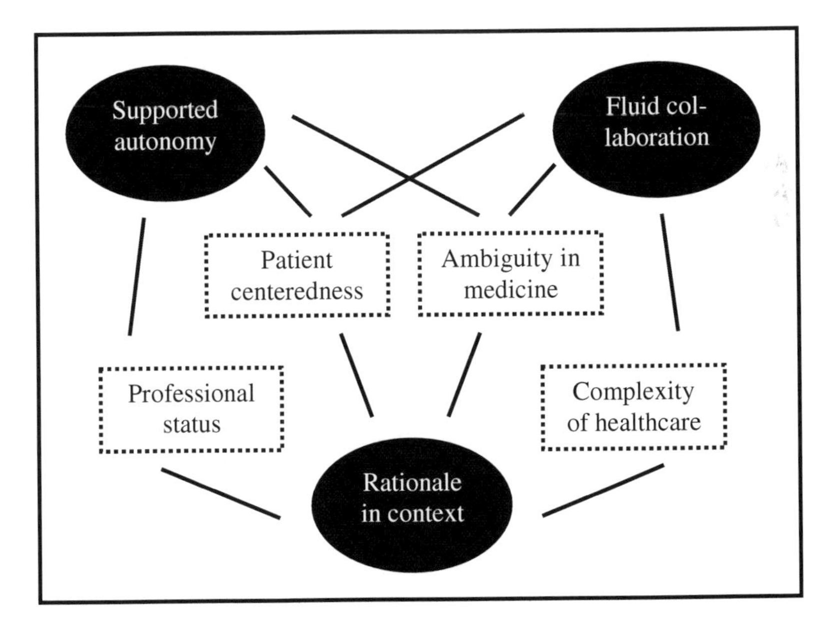


**Fig. 1.** The relationships between the social requirements of health 2.0 applications and the characteristics of collaboration in healthcare

opinions. These differences are important, they are the result of the *patient centeredness* of the healthcare profession: healthcare workers will sometimes disagree because they see different ways to pursue the wellbeing of their patients.

### 4.2 Rationale in Context

The complexity and ambiguity of the medical profession results in a high need for insight in the arguments on which decision of colleagues are based. This means that a healthcare professional needs more than the mere facts, but he or she also needs to get insight in what line of thought that was followed, and by who [20]. That is why, as a second social requirement, we propose 'rationale in context'.

For example, it became clear that standardized documents often successfully fulfil their bureaucratic task, but that a higher standardization could also lead to a reduction of the clinical value of documents, due to the lack of a 'core narrative' [20]. In communication, the 'core narrative', rationale or 'meta communication' serves as a guideline for the party that receives the information and is essential in the recontextualisation process of information. The awareness of activities of colleagues also serves as a source of information and guidance for one's own actions.

**Relationship to the characteristics of collaboration in healthcare:** The *complexity of healthcare* makes that professional healthcare workers often have to rely on information from colleagues. As the medical information is *ambiguous*, it is very important to know who the information comes from, in order to able to interpret the information and fully judge its value. Therefore, they need to understand 'rationale' behind the medical information: to be able to interpret medical information is what separates the layperson from the *professional* healthcare worker. Professional healthcare workers put *patient care first*: they will therefore sometimes disagree with colleagues.

### 4.3 Fluid Collaboration

The last social requirement we propose is 'fluid collaboration'. It is clear that much of the work healthcare professionals do cannot be captured in procedures: much is being done ad hoc and tailored to the patient's needs [12]. Therefore, a system should be as least constraining as possible, as too much constrain will lead to a user boycott and a system failure. This was observed by Trivedi and colleagues [21], who saw that considerable treatment flexibility is necessary in order for a software program to be used in all cases in a real world setting.

This is not to say that workflow should altogether be ignored. On the contrary, the system should be adjusted to the workflow, and prior to implementation, users should evaluate the system for further customisation [21]. When workflow is ignored, a system can impose a 'new reality' and precisely increase the rigidity of the work organisation [22].

However, it is very hard to detect a 'core' workflow, especially in healthcare [18]. So while developers should clearly pay attention to the workflow of the system users, there should always be room for exceptions, and easy adaptations.

**Relationship to the characteristics of collaboration in healthcare:** Healthcare professionals will always put the *patient's wellbeing* before imposed procedures or systems. This makes it hard to formalise procedures in healthcare. The ambiguity of findings further impedes formalisation. However, collaboration is an inevitable part of current work in healthcare, as healthcare has become *highly specialised and complex* and patient will often have to visit several professional healthcare workers.

## 5 Discussion

Collaboration in healthcare is an inevitable challenge. With healthcare becoming more complex and specialised, the need to exchange information, knowledge and practice between healthcare professionals will most likely further increase in the years to come. Health 2.0 applications can be a welcome help in this process, but only if the designers behind these tools also take the social requirements of their users into account. Therefore we have proposed three social requirements: supported autonomy, rationale in context and fluid collaboration.

These are to a certain extent hypothetical and future user research planned in the Share4Health project will serve to validate these social requirements and to further translate them into more detailed recommendations. In order to attain this, we will sit together with technical groups who have explored the possibilities of web 2.0 technology. By combining the technical knowledge with these social requirements, several health 2.0 scenarios will be developed that will further build on the information exchange currently planned within the Share4Health. These scenarios will then be presented to focus groups of GPs and pharmacists for feedback. This should allow us to further specify the social requirements to certain settings in healthcare and profession.

## References


1. O'Reilly, T.: What Is Web 2.0? Design Patterns and Business Models for the Next Generation of Software (2005), `http://www.oreillynet.com/pub/a/oreilly/tim/news/2005/09/30/what-is-web-20.html`
2. Giustini, D.: How Web 2.0 is changing medicine - Is a medical wikipedia the next step? British Medical Journal 333(7582), 1283–1284 (2006)
3. Boulos, M.N.K., Wheeler, S.: The emerging Web 2.0 social software: an enabling suite of sociable technologies in health and health care education. Health Information and Libraries Journal 24(1), 2–23 (2007)
4. Hughes, B., Joshi, I., Wareham, J.: Health 2.0 and Medicine 2.0: tensions and controversies in the field. J. Med. Internet. Res. 10(3), e23 (2008)
5. Eysenbach, G.: Medicine 2.0: Social Networking, Collaboration, Participation, Apomediation, and Openness. J. Med. Internet. Res. 10(3), e22 (2008)
6. Duysburgh, P., Jacobs, A.: Back to the Future of Healthcare:Taking a closer Look at Health 2.0 Discourse (unpublished, 2009)
7. Pilemalm, S., Timpka, T.: Third generation participatory design in health informatics–Making user participation applicable to large-scale information system projects. Journal of Biomedical Informatics 41(2), 327–339 (2008)
8. Grudin, J.: Groupware and social dynamics: eight challenges for developers. Commun. ACM 37(1), 92–105 (1994)

9. Ackerman, M.S.: The intellectual challenge of CSCW: The gap between social requirements and technical feasibility. Human-Computer Interaction 15(2-3), 179–203 (2000)
10. Sanders, E., Stappers, P.J.: Co-creation and the new landscapes of design. CoDesign 4, 5–18 (2008)
11. Harrison, S., Tater, D., Sengers, P.: The Three Paradigms of HCI. In: alt.chi Conference (2007)
12. Schaper, L.K., Pervan, G.P.: ICT and OTs: a model of information and communication technology acceptance and utilisation by occupational therapists. Int. J. Med. Inform. 76(Suppl.1), S212–S221 (2007)
13. England, I., Stewart, D.: Executive management and IT innovation in health: identifying the barriers to adoption. Health Informatics Journal 13(2), 75–87 (2007)
14. McKinlay, J.B., Marceau, L.D.: The end of the golden age of doctoring. International Journal of Health Services 32(2), 379–416 (2002)
15. Kaplan, B.: Evaluating informatics applications - some alternative approaches: theory, social interactionism, and call for methodological pluralism. International Journal of Medical Informatics 64(1), 39–55 (2001)
16. Jensen, T.B., Aanestad, M.: How healthcare professionals "make sense" of an electronic patient record adoption. Information Systems Management 24(1), 29–42 (2007)
17. Reddy, M., et al.: Sociotechnical requirements analysis for clinical systems. Methods of Information in Medicine 42(4), 437–444 (2003)
18. Berg, M.: Patient care information systems and health care work: a sociotechnical approach. International Journal of Medical Informatics 55(2), 87–101 (1999)
19. Dreiseitl, S., Binder, M.: Do physicians value decision support? A look at the effect of decision support systems on physician opinion. Artificial Intelligence in Medicine 33(1), 25–30 (2005)
20. Winthereik, B.R., Vikkelsø, S.: ICT and Integrated Care: Some Dilemmas of Standardising Inter-Organisational Communication. Computer Supported Cooperative Work (CSCW) 14(1), 43–67 (2005)
21. Trivedi, M., et al.: Barriers to implementation of a computerized decision support system for depression: an observational report on lessons learned in real world clinical settings. BMC Medical Informatics and Decision Making 9(1), 6 (2009)
22. Sicotte, C., Denis, J.L., Lehoux, P.: The Computer Based Patient Record: A Strategic Issue in Process Innovation. Journal of Medical Systems 22(6), 431–443 (1998)

# A Web2.0 Platform in Healthcare Created on the Basis of the Real Perceived Need of the Elderly End User

Giovanni Rinaldi, Antonio Gaddi, Arrigo Cicero, Fabio Bonsanto, and Lucio Carnevali

CUP2000, Via del Borgo di S. Pietro 90/c, 40100 Bologna Italy
University of Bologna Medicine Faculty Via Massarenti, 40100 Bologna Italy
giovanni.rinaldi@cup2000.it

**Abstract.** The elderly care is characterized by integration of health and social care, is performed by several different agencies, requires a continuum of assistance and not episodic approach. These considerations has led us to overcome the traditional ICT approach done by several applications connected with property mechanism towards new solutions proposed by web2.0. We think that a federative platform can support the elderly needs (detected through focus groups) providing them web2.0 solutions in health care. The main features of the federation are: publishing/subscribe, messaging and signaling. These dimensions allows elderly to participate actively in their own healthcare through the access to their own health and social record, the composition of their own health and social space, and the participation to social networks for remaining socially active. Moreover, the process of dis-intermediation is also allowed. More emphasis is posed in the federative platform for the governance of the services as requested by elderly.

**Keywords:** Web 2.0, Medicine 2.0, e-Health, personalized healthcare, EHR.

## 1 The Early Approach to Elderly Care and the New Challenges in Web2.0

Among the Italian cities Bologna (Italy) has the highest percentage of elderly above 64 years (26.7%). Families with at least one member over 64 years are 38% of Bologna's families. Families with only one member over 64 years are almost 17% of the families, if we consider only elderly with more than 79 years the same percentage is 7%. Several projects for elderly using ICT have been implemented in the last few years in Italy and especially in Bologna [1]. These projects have been generally called e-Care projects. The goals of the traditional e-Care network are the followings: to connect the stakeholders in the socio-sanitary processes; to share the information during the treatment, to provide a more complete and integrated care using ICT solutions; to integrate health and social services; to collect citizens' health information through the network. These projects consist of several applications aimed to communicate with each others in a peer-to-peer connection, so they tend to build up several data silos with the same data but with different functions. Emphasis is posed on the data exchange and not on the collaboration activity needed for caring people in a

P. Kostkova (Ed.): eHealth 2009, LNICST 27, pp. 173–180, 2010.

multi-agencies environment. In our new project, called OLEDS and co-founded by the EU [2], elderly are provided at home with a simple device managed by a remote control. The system is connected to the TV and to a VoIP telephone; elderly suffering from coronary artery disease are monitored with blutooth easy-to-use telemedicine devices. At this state of the project 30 elderly, 4 volunteers associations, doctors and social workers are experimenting the platform; by the end of the year others 70 elderly will be engaged. The main aims of the project are the followings:

1. To overcome the traditional 1-1 relationship in a one 1-many: in this context we try to promote the aggregation of social networks through the utilization of thematic channel managed by animators;
2. To engage different stakeholder with different responsibilities and associated to different organizations (institutions, volunteers, private sector);
3. To provide a complete care for elderly both on social and health side. For elderly health space cannot be separated form the social space;
4. To manage the "jam of data" collected from elderly: the concept of numerical data in a relational Electronic Health Record now is wider; the information that describes the health or social state of the elderly are composed of different type of data, for instance numerical data coming from devices, videos, images, speech diaries, ...

On the other hand our experiences have led us to modify the approach normally used because we think that the problem is not to propose a particular home devices, but an organizational change driven by ICT and web 2.0. From an organizational point of view:

1. Essentially the elderly care is a multi-agency activity: this commitment doesn't allow to impose solutions it has been done proposing singles vertical application and a middleware layer for the integration; but it requires an activity based on the co-production;
2. The care of the elderly must respond to a complex long term conditions rather than rely on the traditional approach of episodic care; whereas we need to consider the continuum of care.
3. Engagement of elderly and stakeholders in the care processes.
4. The "www" of the elderly is the little but great world of relationships and the main scope of the platform we are going to explain is the representation of this world using the technological benefits of the web2.0.

We have observed the following characteristics of multi-agency care and of the contexts in which shared technical infrastructure and new working practices are being constructed: the participants not only belong to different agencies but also have the different value sets, priorities and perspectives; the policy drivers and management imperatives that bring the parties together often imply or demand second order rather than incremental change; the nature of the health and social care relationships make issues of governance of practice and of information paramount; often the technological systems for care people are "user centric", this is an improvement compared to "function centric approach", but there is no health information out of patient-doctors relationship. We noted that the nature of the problem requires the engagement of several institutional agencies in conjunction with volunteers organizations and the social capital they can bring and the new marketing perspective introduced by private sector requires a new definition of technological system. This new technological system is different from the traditional way considered as several solutions (web or legacy) closely interconnected. For these reasons, inside the project, we have introduced three "technical" terms in response to what we have discovered from contacts with the realities of the provision, delivery and use of service to support elderly people in Bologna:

- The projection orientated approach to systems representation
- Federability approach in designing and delivering services involving multi-agency
- The service orientation to systems organisation

The first one is a response to the need to maintain accessibility and participation in the face of complexity, the last two are a reply to the heterogeneous, multi-agency nature of the context we are engaged. Our commitment to user centered design removes the possibility of segregating policy making, requirements capturing and design spaces. We have to create a space in which these processes co-exist and co-evolve. Responsibilities remain distinct and "boundary objects" are required to mediate and facilitate mutual sense making. This is the purpose of the projections and the way they remain distinct but interrelated at both semiotic and formal levels providing a potential bridge between the languaging and the engineering domains. During the last decade, a large body of projects and technological realizations have been carried out aiming at creating e-Health and telemedicine technologies. There is a relatively low practical application of them in every day practise. One limitation is that, even if theoretically perfect, the realized innovation are far from the perceived need of the final user and from its ability to use them. In this context we have met 70 elderly subjects (mean age: 78+/-6 years old, M:W=1:1.2) in recreational meeting centers in the city of Bologna administrating them a specific questionnaire created to understand their technological skill and their perception about new technologies potentially useful for the management of their health. The preliminary analysis of the questionnaire show that the interviewed elderly are relatively interested in the possibility to use new technologies to manage their health care. In particular they could be interested in technologies that are easy to be used, and could automatically furnish information such as weight, blood pressure or even a basal electrocardiogram. The most part of interviewed are able to use the standard electronic tools such as the phone, television remote control, the kitchen tools and the washing machine. The most part of them measures their blood pressure with hand or automatic sphygmomanometers, but they don't use or don't know the existence of the most common e-health technologies, or they don't surf the internet.

## 2 The Federative Platform as Approach for Web2.0 in Healthcare

The kind of problem described in the previous paragraphs cannot be dealt by using the classical enterprise approach made of property solutions, deployed on the web, closely integrated with proprietary mechanisms. It is because of these challenges: multi-agency solution, continuum of care for elderly, engagement of elderly and stakeholders in the care processes; that we refer to a federal approach which both supports and depends on trust and co-operation between agencies allowing them to maintain their individual relationship with clients and responsibility. In this vision, the processes which maintain, refine and evolve the rules, scripts and practices inscribed in the infrastructural system are seen as an integral part of the use and governance of the supporting information and communications services. A technologically advanced e-Care environment implies a highly distributed collaborative network, composed of heterogeneous and autonomous nodes: all of which are involved in supporting the elderly and delivering or orchestrating services. We are not proposing a conventional application but we are defining a set of services that users may compose in a creative way (within limits imposed by the service logic) to compose their own environment. In the infrastructure we are building, different components (services) may be fit. Services can be developed at any time, when their need arise. Using the conventional «hub and spoke» metaphor, the nucleus of the architecture is the service hub which provides the means of connecting user systems to services systems. The functions which the hub provides are relationship and provisioning services; registration and publication services; event, process and transaction services. The core concept of the OLDES architecture is the service hub which provides the means of connecting user systems to service systems. Additional notions are: (1) Federation is achieved through hub to hub connections and the sharing of third party services; (2) The hub provides the core coordination services which are associated with "middleware", these are: "Portalling" i.e. publication, syndication and organisation of contents; "Switching" i.e. process coordination within and across boundaries ensuring that sequences of events and transactions are managed; "Indexing" i.e. the management of identifiers, tokens, authentications and relationships. (3) The concept of provisioning (consents and capabilities) is extended to apply not simply to role based user access to named resources but also as follows: (a) Services require explicit provisioning to make use of the outputs of other services, (b) Any instance of provisioning involves a link between a source of (sensitive) data, a sink – or user - of the data and the subject of that data: a three way relationship role and relationship based access, (c) The notion of a service session provides the final provisioning concept where sessions are declared and audited by class, e.g. emergency, ongoing case, etc. We think that this architecture can respond to a challenges proposed by web2.0 in health care. In this platform the concept of federation and the associated components allows to elderly to autonomously use technology to access personal information like EH/SR, to manage their personal health, for nurturing social inclusion through social networks. On the other hand, stakeholders belonging to different agencies can participate in the care process in a collaborative way using the tools for composing the services requested by elderly and met on the network through the engine for knowledge search. The key requirement that is being addressed in this architectural approach is one of governability. In care and development environments, information may be accessed and used only on the basis of informed consent and for the purposes intended in the granting of that consent. Not that this is the antithesis of the web2.0 concept of complete absence of accessibility constraints on data about content or use of content. These particular issues of governability require explicit representation at the architectural level because they are an essential part of the requirements and policy discourse. This is an essential characteristic of what in the federative platform we refer to as "service orientation". Issues of capacity, availability and safety may be treated as a technical abstraction and relegated to the negotiation of the service platform provision relationships. In summary then, the OLDES service platform provides a potentially extendable and federable environment for the brokered delivery of services. To better explain we can introduce the following two logical components:

*System Management functions*: functions used for the control of the system.

*Service Management functions*: functions needed to integrate with other hubs, allowing the access of new agencies, the exposition of services and of contents.

Features characterizing the federative platform are:

*"Publishing and Syndication"*: this is the concept of making own information accessible to stakeholders which have necessity to access; in this vision the communication of personal information is an active action based on recognition by the parties of the need to inform and the desire to share and in which the subject of information is an informed and active participant;

*"Messaging"*: in this platform the concept of message does not allow the exchange of data; the message isn't only the vehicle for building or maintaining data base, but it is the way for alerting users also.

*"Signaling"*: it is the observation of the events in the federation. It produces marketing (for profit and no-profit organizations) because the agencies and volunteers organizations can produce new services on the basis of the usage of the services by end users, observed using signaling functions.

The use of data in the federative platform is worth a further discussion. The platform needs not only data produced within its boundaries but also data that can be produced and owned by other agencies within the federation. A specific service (or set of services) must be responsible of recovering information from different repositories belonging to different agencies and compose them in a transparent way on the user interface of the care-giver. From a logical point of view, we have to introduce a service for data management which scope is not to aggregate every type of information in a single data base, but to extract information and compose a specific record from federated databases. The notion of governance which has been used so far applies to the definition, commissioning, delivery and evaluation of service. It is concerned with the means and the mechanisms by which authorized personnel is able to ensure that the service environment is directed and operates in an appropriate and effective way. Within this concept, we can distinguish a further aspect or level of governance which applies to the information that is generated and interpreted in service processes, particularly when they are mediated or supported by electronic means. The importance of this distinction in the governance rests on the idea that there are information and communications services which are shared by a number of higher level, care services, and, through sharing these underlying services, higher level services are able to co-ordinate their operations. This continuing independence and distinctive identity of the participants means that the term "integration" is no longer appropriate. This coordination of independent entities, through the sharing of (third party) information and communications services, is the core concept of Federability and, as such, can be seen to be closely associated with the sort of service oriented approach we are describing here. This leads us to the imperative we face: agencies within clinical and social care, the voluntary sectors and the commercial sector are, and will remain, independent but only if they co-ordinate and co-operate that the gaps between needs and available resources in the care of the elderly can be addressed. The core purpose of this service infrastructure is to make this co-ordination possible [4].

## 3 The Contribution to Web 2.0 in Healthcare for Elderly

As a contribution to Web2.0 in healthcare we think that the implemented platform can reply to the request of providing to the patients some tools for promoting their participation in the care process. In particular, taking [3] as a starting point we introduce some dimensions that describe how some functions can provide the autonomy to the patient in his "collaboration" with stakeholder to improve their own wellbeing. The emphasis is posed on the autonomy of the patient in searching and using health services on the net. The experiences done has demonstrated that when we design ICT system for health using the capabilities of the internet we have to keep in mind the user target and the method for facilitating the entrance of the technology in patients' houses and the engagement of different agencies involved in the care process. The first feature we would like to analyze is the EHR built by professionals but accessible by patients. In this context, we prefer to refer to an Electronic Health and Social Record (EH/SR) for stressing the concept of collaboration among different agencies discussed above and the concept of elderly wellbeing that does not separate the health activities from the social ones. In this context we consider an EH/SR personally controlled by elderly, and build by professionals and by elderly themselves as a kind of simplified health space. The main areas of the personal health space are:

1. Questionnaires for specific communication with professionals. In this area elderly are driven to fill suitable questionnaires about social state or about specific chronic disease.
2. Observational and narrative speech diary. These features give to the elderly an opportunity to record a free speech diary about their health problems or social anxieties.
3. Visualization of data coming from devices; data coming from telemedicine devices are visualized and presented to elderly.
4. Personal booking and appointment; elderly with the help of volunteers, record and remember personal appointments.

Of course, professionals complete the areas of the EH/SR of the competence of doctors and social workers. Another area of participation proposed by web2.0 is the opportunity for elderly to participate to social networks. In the context of this platform social networks are thematic channels proposed by animators that engage elderly in conversations or entertainment. The contribution of different agencies in the organization of thematic channel is an important challenge to the utilization of the platform by elderly. An interesting feature proposed in [5] describes a special form of disintermediation where people who search health information on the web receive the guide from "apomediaries"; they are agents (people or tools) which "stand-by" and guide the users to services. It is noted in [6] that the process of disintermediation can follow a route from intermediation to disintermediation when the user reaches a sufficient autonomy, but people can return to intermediation, so the relationship with intermediary is not ended. The federative implemented platform can be considered a mix of brokerage and disintermediation and is stressed the dynamic previously discussed.

The elderly expresses a strong request for socialization and this is provided by web technologies. In this platform apomediaries are really the volunteer associations and the automatic tools that help elderly to extend their trust relationship and help elderly to obtain the specific information they need (the access to the health space, the request of services). But in this mechanism we have to consider that EH/SR in the platform is centered on the elderly-doctors relationship. This implies a trust relationship, so the intermediation-disintermediation process needs both a brokerage activity performed both by tools and by intermediaries for guiding elderly and stakeholders to the choice of the request information. On the other hand for the stakeholders the process of analysis of information produced inside the platform and the search of information is very important [7]; but we would like to note that the different information exchanged also produce a sort of marketing where stakeholders can generate new services and new thematic channel on the basis of the real requirements of the elderly. In fact at the core of this disintermediation tool is an advanced profiling, matching and targeting system that will allow to continually enhance the relevance of the content offered to the targeted audiences, the degree of their usage and participation, as well as the efficiency and effectiveness of the thematic channel production, moderation, assistance, support and system availability processes. This system is based on contextual, behavioral and social networking targeting, making use of knowledge/intelligence extracted automatically from unstructured content (thematic channels, speech diary, communication between all the stakeholders of the system, questionnaires), from historical analysis of user – system interactions and from sensors data received from the medical sensors. The extracted knowledge along with other specific enterprise knowledge (knowledge contributed by the experts of the federative platform; definition of alarms, of triggers, of processes, best practices, etc.) is maintained in a dedicated Knowledge Base. Within the federative platform, all the interactions between the stakeholders are formalised in well defined models (events, tasks, processes, best practices) that will take advantage of the technologies developed and allows the presentation of specific scenarios. For this reason, the mechanisms of signaling and logging play an important role because they allow the generation of elderly profile and the health scenarios used by stakeholder for better care the patient or for proposing the services requested.

## 4 Conclusions

In several years of elderly care projects we have learned that the elderly care cannot be addressed through en episodic care approach, it must be a continuing assistance both in the health care and in the social world. The concept of wellbeing that we are pursuing does not allow these two aspects of care to be separated and treated independently. This is more clear if we think that the care of elderly is done by health institutions and social workers; and the "little" world of elderly is done by the trust relationships with doctors, social workers, family, friends, parishes, volunteers organizations; and a real system of care cannot exclude one of these components of the elderly life. For this reason we thought to work about the organizational change necessary for responding to this understanding. This led us to search the novelty brought by web 2.0 and we proposed a federative platform with the features of publishing and subscribe, messaging and signalling. In this context, the platform responds to the requests of elderly (recognized from the outset of the project through focus groups) by providing an infrastructure capable of supporting the operation and governance of a dynamic and participative e-care, and this implies a requirements for regulation, governance, consent or confidentiality, and the important issues of governability is an essential part of the requirements and policies commitment. The federative platform uses a specific set of web tools for responding to the request of generation of content by users and the power of the platform is used in order to personalize health care, collaborate, and promote health education such as connecting elderly with same interests or chronic disease and improving an individual's value from health and social care. The EH/SR is built by professionals, but is accessible by elderly; and nearby a sort of health space is built up where elderly can record a speech diary and where health and social questionnaires are proposed. In the platform, the usage of the data allows a more strict collaboration with stakeholders because information is not a property features of a data silos. Moreover the platform allows the participation to social network animated by volunteer; this allows elderly to be social active and in relationship with other people. The mechanism of generation of new contents, services and social networks is composed using the automatic tools that analyze the requests performed by elderly. In this context of dis-intermediation elderly can personalized own health and animators can compose their new services. On stakeholders' side the data gathered (questionnaires, usage of the services, data from telemedicine devices, speech diary, ...) will be treated by automatic tools for composing elderly profile and for proposing scenarios for better care patients.

## References

1. Rinaldi, G., Lena, C., Tomba, R.: La presa in carico elettronica del cittadino nel processo di assistenza; l'esperienza eCare nell'area metropolitana di Bologna. Salute e Società. Anno II 2/2003, 93–114 (2003)
2. OLDES Project, http://www.oldes.eu
3. Giustini, D.: How Web 2.0 is changing medicine. BMJ 2006 333(7582), 1283–1284 (2006)
4. Aas, M.: The organizational challenge for health care from telemedicine and eHealth. Arbeidsforskningsinstituttet AS (2007) ISBN 978-827609-208-0, http://www.afi.no/stream_file.asp?iEntityId=2088
5. Eysenbach, G.: Medicine 2.0: Social Networking, Collaboration, Participation, Apomediation, and Openness. J. Med. Int. Research 10(3), e22 (2008)
6. Eysenbach, G.: Credibility of health information and digital media: new perspectives and implication for youth. In: Metzeger, M.J., Flanagin, A.J. (eds.) Digital Media, Youth, and Credibility. The J D and Catherine T MacArthur Foundation Series on Digital Media and Learning. MIT Press, Cambridge (2008)
7. Thelwall, M.: Extracting accurate and complete results from search engines: case study Windows Live. J. Am. Soc. Inf. Sci.Technol. 59(1) (2008)

# Web 2.0 Artifacts in Healthcare Management Portals-State-of-the-Art in German Health Care Companies

Nadine Blinn, Mirko Kühne, and Markus Nüttgens

University of Hamburg, School of Business, Economics and Social Sciences
Von-Melle-Park 9, 20146 Hamburg, Germany
{nadine.blinn,mirko.kuehne,markus.nuettgens}@wiso.uni-hamburg.de

**Abstract.** The internet is increasingly used as a source for information and knowledge. Even in the field of healthcare, information is widely available in the internet. In the context of healthcare management, two general questions are of interest: which information or content is provided and how is provided by whom? As sickness funds play a highly relevant role in the German healthcare system, we conduct an exploratory survey to answer these questions. We perform a third party web assessment by doing a complete inventory count of the German sickness funds landscape. Our study provides a foundation for further research by raising first categories that can be used for a theoretical explanatory model.

**Keywords:** Web 2.0 artifacts, German healthcare system, complete inventory count.

## 1 Introduction

The Internet is increasingly used as a source of information and knowledge. Since people are more and more used to gather information about a wide variety of topics by electronic means, nearly all thematic aspects of daily life are covered. Hence, information that addresses the field of healthcare in general or specific health care aspects (e. g. specific diseases) is increasingly searched online [1] [2]. In the German health care system, sickness funds play a highly relevant role as they provide most of the publicly funded healthcare system on the national level. Because of recent developments in German health policy, sickness funds are under high cost pressure. Moreover, competition between today's 238 sickness funds is very high: as contributions of insurants in Germany were harmonized to a certain level, the differentiating factors for the companies are the services they provide. Consequently, German sickness funds have two main reasons to provide extensive and high quality information via Internet:

- Cost savings: the provision of detailed, structured and extensive information to the insurants aims at avoiding costs for personal, time-consuming consultation. Therefore providing information by websites is cheaper than providing telephone customer services or personal services in an agency.

P. Kostkova (Ed.): eHealth 2009, LNICST 27, pp. 181–188, 2010.

- Competitiveness: as the fee for health insurances in Germany was harmonized to a uniform level, sickness funds need other major differentiating factors from their competitors than costs. One possibility that can also be found as a major differentiating factor in other industries is "quality of services". Consequently, information services provided by electronic means with a wide availability and a certain quality level could be a factor.

Because of the increasing use of the Internet as a source of healthcare knowledge and information on the one hand and the special role and competitive situation of sickness funds in Germany on the other hand, we expect them to provide high quality websites that offer a wide variety of information. Therefore, we formulate two initial research questions: (1) by what amount do Germans sickness funds provide information to insurants and interested persons? (2) By which technologies are the services provided? The goal of this paper is to provide first exploratory findings concerning these two questions from a survey of websites provided by German health insurance companies. We conducted the survey according to the methodology "third party web assessment" [3] and performed a complete inventory count comprising websites of all German sickness funds. In order to get objective – or at least intersubjectively verifiable – measurements for the provided content as well as for the used technologies, we use two different benchmarks:

(1) For the contents, we use the so called "healthcare bulb" [4] a model comprising different layers with detailed criteria to cover all thematic clusters belonging to healthcare management.
(2) For the used technologies, we refer to a framework for Web 2.0 characteristics [5] [6] in order to structure the technological perspective.

The paper is structured as follows. In the second section, we depict the background of the German healthcare management system. We complement the basic principles of national healthcare system by an overview on related work. In the third section, we present our study. After having depicted the design and methodology of the study, we present and discuss the results in a detailed manner. The article closes with a summary and an outlook on further research.

## 2 Basic Principles

### 2.1 The Landscape of German Healthcare Management

The German health insurance reform of 2007 requires everyone living in Germany to be insured [7]. There are two main types of health insurance – the *public health insurance* ("Gesetzliche Krankenversicherung" or GKV), which is also known as *sickness funds*, and the *private health insurance* ("Private Krankenversicherung" or PKV). Approximately 85% of the population is member of the one of the public sickness funds, while the others usually have private health insurance. Consequently, most German residents (approx. 70 million people) are insured by the public system (except public officers, self-employed people and employees with a gross income above 48.600 EUR per year or 4,050 Euros per month [8].

As required by law within the fifth social statute book, (SGB V) members of the sickness funds have to pay an insurance fee which depends on the amount on their income as employees. Each insurant gets the same benefits, even though the individually paid fees might differ. All of the sickness funds must charge the same rate. Consequently, there is no longer any competition between sickness funds based on fees; competition only exists on services and possible refunds. Usually, the company of an insured employee pays half of the insurance contributions, the other half is provided out of the employee's salary. The fee for this public health insurance is currently 15.5% (from July 1, 2009: 14. 9%) of the eligible gross salary to a maximum monthly income limit of 3,675 EUR. The previously large number of public health insurances decreases every year: currently the Federal Ministry for Health officially counts 196 [8].

The private insurance system is based on an individual agreement between the insurance company and the customer. The fee depends on a range of individual characteristics, for example, the percentage of coverage, the amount of chosen services, the individual risk or the entrance age into the private system, and so forth. The private health insurance market is well served by about 50 German insurance companies [9]. Both the public health insurance and the private health insurance struggle with the increasing cost of medical treatment and the changing demography [10].

### 2.2 Related Work

In the research field of electronic health, several studies exist addressing how German health insurances use the Internet for their business' purposes or how Web 2.0 technologies and concepts are used in the context of healthcare [1; 11; 12; 13; 14] However, to our knowledge, none of these give a complete and detailed overview about which content is provided by sickness funds' websites or which technologies are applied for presenting the content.

## 3 Survey of German Health Insurance Companies Websites

### 3.1 Methodology and Design

The accomplishment of our study follows the method of "third-party web assessment" [3] whereas the "mystery user" approach is applied [15]. The principle of the "mystery user" approach indicates that an examiner puts her- or himself in the role of a client that requires the services provided by the website. This methodological approach is also known as "mystery shopping" [16]. In order to benchmark the examined websites, we apply the healthcare bulb on the one hand to evaluate the provided content and a framework for Web 2.0 characteristics on the other hand for measuring the technological perspective. As we conducted a complete inventory count, the database comprises all 46 German PKV[1] and 192[2] German GKV 238 data

[1] For an overview compare: http://www.pkv.de/verband/mitgliedsunternehmen/
[2] For an overview compare: *https://www.gkv-spitzenverband.de/publish/Alle_gesetzlichen_Krankenkassen.aspx?ActiveID=1028&gvAdressenOverview_PageIX=0* and *http://www.krankenkassen.de/gesetzliche-krankenkassen/krankenkassen-liste/*

sets are gathered in total. The criteria catalogue for the content consists of 21 criteria (cp. Section 3.1.1) and the catalogue for the technical perspective consists of five criteria (cp. Section 3.1.2). The evaluated 26 criteria are transformed to a binary scoring model, that is, if a criterion is fulfilled (information for a criterion according to the healthcare bulb is available), the portal gets one point and otherwise it gets zero points. By aggregating the points over all criteria, the maximum number of reachable points is 26.

#### 3.1.1 Content Benchmark

In general, a common sense of what the term "health" means is intuitively given. While the World Health Organization defines health as "a state of complete physical, mental and social and well-being and not merely the absence of disease or infirmity" [17], the literature does not give a consistent definition [18]. Therefore, we cannot clearly determine the characteristics of "healthcare", or what "healthcare" really and objectively is [19]. The "healthcare bulb", a model consisting of four layers, provides a possibility to structure the components or stakeholders of healthcare management. The healthcare bulb exists in different variations [19][20][21]; however, all variations refer to the initial model of [4]. Accordingly, health care management comprises – comparable to a bulbs' layers – core areas as ambulatory and clinical healthcare treatment services. The core areas are complemented by pre-service industries and supplying industries as well as border area industries and other branches with strong relations to healthcare [4].

#### 3.1.2 Technological Benchmark

The use of Web 2.0 applications in a medical context or in the context of healthcare is defined as "web-based services for health care consumers, caregivers, patients, health professionals, and biomedical researchers, that use Web 2.0 technologies and/or semantic web and virtual reality approaches to enable and facilitate specifically 1) social networking, 2) participation, 3) apomediation, 4) openness, and 5) collaboration, within and between these user groups [22]. This principle is also called "Medicine 2.0“, "Health 2.0" or "e-health" [1] [14] [22]. Within our survey and according to our methodology, the patients' and health care consumers' view in their role as a user is focused. Consequently, the possibilities of Web 2.0 applications from the users' point of view can be distinguished according to [23] within: Authoring, Sharing, Collaboration, Networking, as well as Scoring. The corresponding applications and principles are Weblog, Wiki, Social Tagging/Social Bookmarking, Social Networking and Podcasts.

#### 3.1.3 Overall Criteria Catalogue

According to the specifications in section 3.1.2 and 3.1.3, the criteria catalogue presented in Table 3 provides the categories which we use for our exploratory survey.

**Table 3.** Criteria Catalogue

| Category | Healthcare Bulb Layer | Criteria |
|---|---|---|
| Content | 1<br>Ambulatory and clinical healthcare treatment services as well as care | Clinic / Hospital, Independent health practitioner, Alternative medicine, Rehabilitation, Care facility |
| | 2<br>Healthcare administration | Administration, Pharmacy, Cure- and bath resorts, Self-help |
| | 3<br>Pre-service and supplying industries | Medicine- and gerontology technology, Bio- and gene technology, Health products, Healthtrade, Pharma, Consulting (B2B) |
| | 4<br>Boundary and neighbour area | Sport and leisure, Training and education. Wellness, Nutrition, Living, Tourism |
| Tech-nology | | Blog, Wiki, Social Tagging, Social Networking, Podcast |

### 3.2 Results

In the following section, we present the descriptive results, sectioned according to results for the GKV and PKV.

#### 3.2.1 Results for GKV (Content)

A remarkable finding is that none of the 192 analysed websites fulfils all the defined criteria. Three websites share the first place. They contain most of the content and were evaluated with 18 of 21 points (86%). The second place with 17 points (81%) is shared by the websites of eleven companies. The overall average is 10,3 points (49%) per company. Concerning the fulfilled content of a certain category, the categories *Administration* and *Consulting (b2b)* score the best with 98% and 95% (see Figure 3). The second ranked category is *Self-help* with 91%.The next major group is a bundle of six criteria with a strong focus on personal education, well-being and recreation. It starts with S*port and Leisure (86%)* and *Training and Education (79%).* In the range between 73% and 71% are located: *Rehabilitation, Alternative Medicine, Cure- and Bath Resorts* and *Wellness.* The categories focusing technological aspects such as *Healthtrade* as well as *Medicine- and Gerontology Techniques* are not fulfilled at all, and reach no points.

#### 3.2.2 Results for GKV (Technology)

The results concerning the Web 2.0 artifacts are as allows: 34,38% of GKV have Web 2.0 technologies implemented on their web sites. The most used Web 2.0 artifact is *Wiki* (17,71%). 14,06 % of GKV websites offer a blog, and 12,5% offer podcasts to the users. 10,42 % of the GKV websites provide a social community. Only 1,04 % of the websites provide the possibility for social tagging. With regard to the sum of implemented artifacts per GKV, 54, 55 % implement one technology, 33,33 % use two artifacts and 13 % offer three or more to the users. None of the GKVs offer all five artifacts.

#### 3.2.3 Results for PKV (Content)

None of the 46 analysed websites fulfils all the defined criteria (see Figure 5). The website, that contains most of the content, is evaluated with 16 of 21 points (76%). The second place with 14 points (67%) is shared by the websites of three compa-nies. The overall average is 7,7 points (36%) per company. The content of the categories *Administration* and *Consulting (b2b)* is represented by 98% of the websites. The second ranked category is *Training and Education* with 87%. *Sport and Leisure, Wellness* and *Cure- and Bath Resorts* are following with 54% to 57%. The categories focusing technological aspects such as *Healthtrade, Bio- and Gene Technology* as well as *Medicine- and Gerontology Technology* are underrepresented with a maximum of 4%. Moreover, the information concerning *Pharma* is very little for all websites.

#### 3.2.4 Results for PKV (Technology)

A remarkable finding is that only 2 of the 46 PKV use Web 2.0 artifacts – one insurer provides a Blog and one insurer provides a wiki.

### 3.3 Discussion

In total, the websites of the public health insurance companies reached in total more points than the websites of the private health insurance companies. The public insurances contain in the first, second and fourth layer of the healthcare bulb more content than the private insurance companies. Moreover, the number of points of the public health insurance websites in the first layer is almost twice as high as the number of points of the private health insurance websites. Furthermore, the public websites contain 20% more content on the second layer. The third layer is almost similar – the private companies score 6% more points. In the fourth layer, the public companies cover with a total of 54% approximately 10% more content than the private companies. Overall, the websites of the public companies cover in 15 categories more content than the websites of the private companies. However, the private companies reached significantly more points in the category health products than the public companies. With regard to our initial research questions, healthcare insurance companies focus on content from categories such as Administration, Consultation and Self-help. Concerning the technological perspective, healthcare insurance companies employ various Web 2.0 artifacts. However, our findings also suggest that a significant difference exists (1) between the content provided by GKV and PKV, and (2) between the Web 2.0 artifacts applied by GKV and PKV respectively. Because of these results, we point to two very interesting open issues which remain for further research:

(1) How can the differences between PKV and GKV in the amount of content be explained? Which external factors might provide an explanation?
(2) Why do PKV hardly use Web 2.0 artifacts? Do they use other channels to provide information to their insurants?

Moreover, we are interested if we can use our exploratory results and the used categories as a first starting point in order to develop a theoretical explanatory model for this situation.

## 4 Summary and Outlook

In the presented paper, we aimed at answering two questions: (1) which information or content is provided by German sickness funds and (2) how is it provided? As sickness funds play a highly relevant role in the German healthcare system, we conducted an exploratory survey to answer these questions. Our findings show that the presented amount of content by public healthcare insurance companies (GKV) is higher than the presented amount of content by private healthcare insurance companies (PKV). The same applies to the implementation of Web 2.0 artifacts. There is a need for theory to be developed and further research is needed. Our categories might provide a foundation for a maturity model for health care management websites. Moreover, they provide the starting point for developing a rigorous explanatory model.

## References


1. Kummervold, P., Chronaki, C., Lausen, B., Prokosch, H.-U., Rasmussen, J., Santana, S., Staniszewski, A., Wangberg, S.: eHealth Trends in Europe 2005-2007: A Population-Based Survey. Journal of Medical Internet Research 10(4), e42 (2008)
2. Satterlund, M., McCaul, K., Sandgren, A.: Information Gathering Over Time by Breast Cancer Patients. Journal of Medical Internet Research 5(3), e15 (2003)
3. Irani, Z., Love, P.: Evaluating Information Systems: Public and Private Sector. Butterworth-Heinemann, Butterworths (2008)
4. Hilbert, J., Fretschner, R., Meier, B., Borchers, U., Heinze, R.: Gesundheitswirtschaft in OstWestfalenLippe – Stärken, Chancen und Gestaltungsmöglichkeiten. Bochum (2003)
5. Kolo, C., Eichner, D.: Web 2.0 und der neue Internet-Boom – Was ist es, was treibt es und was bedeutet es für Unternehmen (2006), `http://www.robertundhorst.de/v2/img/%20downloads/Web_2.0.pdf?PHPSESSID=3b45d404f7fec55a20ce077e2b7c6ab2` (Last access: January 2, 2009)
6. Duschinski, H.: Web 2.0 – Chancen und Risiken für die Unternehmenskommunikation, Diplomica, Hamburg (2007)
7. Bundesrat: Gesetz zur Stärkung des Wettbewerbs in der gesetzlichen Krankenver sicherung (GKV- Wettbewerbsstärkungsgesetz-GKV-WSG). In: Bundesrat (ed.) Gesetzesbeschluss des Deutschen Bundestages, Drucksache 75/07 (2007)
8. Bundesministerium für Gesundheit (2009) Gesetzliche Krankenversicherung – Mitglieder, mitversicherte Angehörige, Beitragssätze und Krankenstand; Ergebnisse der GKV-Statistik KM1 (2009)
9. Verband der privaten Krankenkassen (2008): Zahlenbericht der privaten Krankenversicherung 2007/2008, Köln (2008)
10. European Observatory in Healthcare Systems (2000): Health Care Systems in Transition – Germany (2000)
11. PricewaterhouseCoopers Unternehmensberatung GmbH (2001): Gesundheitsportale 2001, Düsseldorf (2001)
12. AMC Assekuranz Marketing Circle (2008a): Assekuranzen im Internet. Alle Versicherungs-Websites im Vergleich. Düsseldorf (2008)
13. AMC Assekuranz Marketing Circle (2008b): Web 2.0 in der Assekuranz – Anwendungsmöglichkeiten, Praxisbericht eund Trends. Düsseldorf (2008)

14. Kovic, I., Lulic, I., Brumini, G.: Examining the Medical Blogosphere: An Online Survey of Medical Bloggers. Journal of Medical Internet Research 2008 10(3), e28 (2008)
15. Heeks, R.: Benchmarking eGovernment. Improving the National and International Measurement, Evaluation and Comparison of eGovernment. iGovernment Working Paper Series. In: Institute for Development Policy and Management (ed.), University of Manchester England (2006)
16. Wilson, A.: The role of mystery shopping in service performance. Managing Service Quality (6:8), 414–420 (1998)
17. World Health Organization (1948): Preamble to the Constitution of the World Health Organization as adopted by the International Health Conference, New York, June 19-22 (1946); signed on July 22 1946 by the representatives of 61 States (Official Records of the World Health Organization, no. 2, p. 100) and entered into force on April 7 (1948)
18. Franck, A.: Der Gesundheitsbegriff des Jedermanns: Studien zum Wandel des Gesundheitsbegriffs anhand der deutschen Literatur vom Mittelalter bis heute. Dissertation Marburg (2007)
19. Dahlbeck, E., Hilbert, J., Potratz, J.: Gesundheitsregionen im Vergleich – Auf der Suche nach Erfolgstrategien. In: Institut. Arbeit und Technik Jahrbuch 2003/2004, pp. 82–102 (2004)
20. Busse, G., Finke, F.-P.: Prospect II – Branchenexposé Gesundheitswirtschaft OWL. Detmold (2005)
21. Hilbert, J.: Gesundheitswirtschaft - Innovationen für mehr Lebensqualität als Motor für Arbeit und Wettbewerbsfähigkeit. In: Institut Arbeit und Technik (ed.) Jahrbuch 2007, pp. 1–16 (2007)
22. Eysenbach, G.: Medicine 2.0: Social Networking, Collaboration, Participation, Apomediation, and Openness. Journal of Medical Internet Research 10(3), e22 (2008)
23. Pleil, T.: Social Software im Redaktionsmarketing (2006), `http://thomaspleil.files.wordpress.com/2006/09/pleil-medien-2-0.pdf` (Last access: January 2, 2009)

# Compensation of Handicap and Autonomy Loss through e-Technologies and Home Automation for Elderly People in Rural Regions: An Actual Need for International Initiatives Networks

Laurent Billonnet[1], Jean-Michel Dumas[1], Emmanuel Desbordes[2], and Bertrand Lapôtre[3]

[1] University of Limoges / ENSIL - Parc ESTER - 87068 Limoges Cedex – France
laurent.billonnet@unilim.fr, dumas@ensil.unilim.fr
[2] Lycée Jean Favard - 27 route de Courtille - 23000 Guéret – France
emmanuel.desbordes@ac-limoges.fr
[3] CCGSV - 9, avenue Charles de Gaulle - 23006 Guéret Cedex – France
bertrand.lapotre@cc-gueret.fr

**Abstract.** To face the problems of elderly and disabled people in a rural environment, the district of Guéret (department of Creuse, France) has set up the "Home automation and Health Pole". In association with the University of Limoges, this structure is based on the use of e-technologies together with home automation techniques. In this frame, many international collaborations attempts have started through a BSc diploma. This paper sums up these different collaborations and directions.

**Keywords:** Home automation, elderly people, e-technologies, international collaboration, universities exchanges.

## 1 Introduction

The department of Creuse today prefigures how large areas of Europe will stand within the next 20 years. The rural population of the department of Creuse is at now one of the oldest in Europe. To take benefit of this demographic reality, the district of Guéret [1], (19 towns for 29 000 inhabitants) decided, 3 years ago, to create the "Home automation and Health Pole". The aim is to drive a coherent action plan in terms of comfort, safety, autonomy and communication for the elderly and disabled people. In conjunction with this, the University of Limoges now proposes for its first year a BSc dedicated to "Home automation and people autonomy". The last ongoing extensions are now necessarily made via international relationships thru universities exchanges and European network of Living-Labs (EnoLL) program [2].

## 2 The Local Initiative Context

E-communication techniques and networks technology together with home automation management constitute the backbone of this rural healthcare initiative.

P. Kostkova (Ed.): eHealth 2009, LNICST 27, pp. 189–191, 2010.

It has to answer and fit the following objectives:

- Improvement of the living conditions of elderly people, in respect for their socio-economic environment, following ethical rules;
- Help for the development of home automation companies and associated services for health and assistance domains;
- Participation of the university and the education system in specific training and research programs.

Concerning this last item, a new BSc diploma has been created and has welcome its first students in September 2008. This diploma has been created jointly with the University of Limoges and the district of Guéret.

More recently, different public and private partners have joined in the EnoLL program and as been validated as a "Living-Lab", thus emphasizing the need for coordinating actions in the domain:

- The region (Limousin) and department (Creuse) councils;
- The hospital center of Guéret;
- The Legrand industrial group, world leader in home automation equipments;
- Axione, the regional telecommunications operator;
- The Chamber of Commerce and professional offices;
- A wide range of official organizations, foundations and medical centers which services are dedicated to elderly and disabled people;
- The University of Limoges through teaching and research programs:
  - National research laboratories working on telecommunications and e-technologies such as XLIM [3];
  - The engineering school ENSIL [4];
  - The Science and Technique college with the BSc "Home automation and Autonomy";
  - The University hospital center.

Applied experiments are today under deployment and preliminary results can now be reported on:

- Field trials: supply of 10 show apartments, communication networks optimization (digital subscriber line - ADSL, optical fiber, power lines -PLC, radiofrequency base stations, …);
- Experiments of home automation products in the hospital of Guéret and surrounding medical centers;
- Economic effects;
- First results of the work placements of the BSc program students;
- Impact on the medical staff and people. Adaptation / resistance to new practices monitored by e-technology. Inferred ethical problems.

Scientific and technical experiments at a research level are also in process:

- Indoor communication studies in a perturbed environment (hospital);
- Intelligent distributed home automation/communication network for daily activities assistance and monitoring;

- Intelligent floor for people activity monitoring, people localization inside the house and real-time dangerous situation alarms;
- Intelligent multimedia real-time network share notebook (unique and centralized contextual multimedia tool for the disabled people, the family, the medical doctors and assistants, the home service assistants ...);
- Evaluation of the disabled or elderly people in terms of ergonomy, mental capabilities in conjunction with home automation equipments.

The "Home automation and Health Pole" of Guéret has got several national and European awards and now has the objective, in association with, and through the University of Limoges researchers experience, to initiate international collaborations.

Among these is scheduled a collaboration with the DOMUS laboratory of the University of Sherbrooke in Quebec, Canada. The DOMUS laboratory is a multidisciplinary lab aimed at research in DOmotics and Mobile Computer Science and located at Université de Sherbrooke [5].

Another one is within the frame of the Norwegian Åsgard program [6] which objectives are:

- To stimulate exchanges between France and Norway in terms of the expertise and scientific competences;
- To emphasize the initiatives of cooperation in research & development;
- To enable French and Norwegian researchers to extend their own international networks.

## References

1. http://www.cc-gueret.fr/
2. European Network Of Living Labs, http://www.openlivinglabs.eu/
3. XLIM Research Institute, http://www.xlim.fr/en/
4. ENSIL Engineering School of Limoges, http://www.ensil.fr/
5. DOMUS laboratory in Sherbrooke, Canada, http://domus.usherbrooke.ca/?locale=en
6. Åsgard franco-norwegian program, http://www.france.no/pages/sciences/asgard_fr.html

# Modeling Market Shares of Competing (e)Care Providers

Jan Van Ooteghem, Tom Tesch, Sofie Verbrugge, Ann Ackaert, Didier Colle, Mario Pickavet, and Piet Demeester

Ghent University – IBBT, Dept. of Information Technology (INTEC),
G. Crommenlaan 8 bus 201, 9050 Gent, Belgium
Tel.: +32 9 33 14900, fax: +32 9 33 14899
{jan.vanooteghem,et al.}@intec.ugent.be

**Abstract.** In order to address the increasing costs of providing care to the growing group of elderly, efficiency gains through eCare solutions seem an obvious solution. Unfortunately not many techno-economic business models to evaluate the return of these investments are available. The construction of a business case for care for the elderly as they move through different levels of dependency and the effect of introducing an eCare service, is the intended application of the model. The simulation model presented in this paper allows for modeling evolution of market shares of competing care providers. Four tiers are defined, based on the dependency level of the elderly, for which the market shares are determined. The model takes into account available capacity of the different care providers, in- and outflow distribution between tiers and churn between providers within tiers.

**Keywords:** eHealth, Techno-economics, Churn, Capacity modeling.

## 1 Introduction

Most OECD countries, particularly those in the EU and Japan, will have to face the challenges raised by their ageing populations in the next decades [1]. It is a widely held belief that the introduction of ICT based eHealth initiatives to improve the efficiency of care provisioning could lead to both a reduction in cost and an improvement in overall standard of living for the elderly. For several years many initiatives have been taken to explore the technical and social implications of a wide variety of such eCare schemes [2][3]. Still large-scale rollouts have been lacking due to the absence of convincing business models.

The work presented in this paper is part of the IBBT TranseCare project [4], a research project aimed at developing ICT support tools and services for an eCare platform. It encompasses diverse fields including social, technical, economic and legal contexts. A custom configurable eCare platform has been developed within the project that connects to the users' television through the Internet, and allows for videotelephony with a professional care help-desk, other users of the eCare system and informal caregivers using an instant messaging client on their PC.

P. Kostkova (Ed.): eHealth 2009, LNICST 27, pp. 192–199, 2010.

Within the IBBT TranseCare project a tool for techno-economic evaluation of the envisaged eCare platform will be elaborated, aimed at to value network [5] and multi-actor analysis. This paper is a first step in this work, and focuses on market share forecasting for competing care services. The model takes into account different levels of dependency (and required care) of the elderly population. Estimating the influence of demographic evolution and current/future capacity limitations in high-dependency care provision facilities can help formulating an eCare case towards governmental agencies, insurance companies or private investors as to how much and how soon to invest in eHealth.

The paper first introduces the conceptual framework of the model and provides definitions for the used concepts. Next an intuitive overview is given of how the model functions, followed by simulation results based on the Flemish situation. Finally we end this paper with some conclusions and future work.

## 2 Conceptual Framework and Model Mechanisms

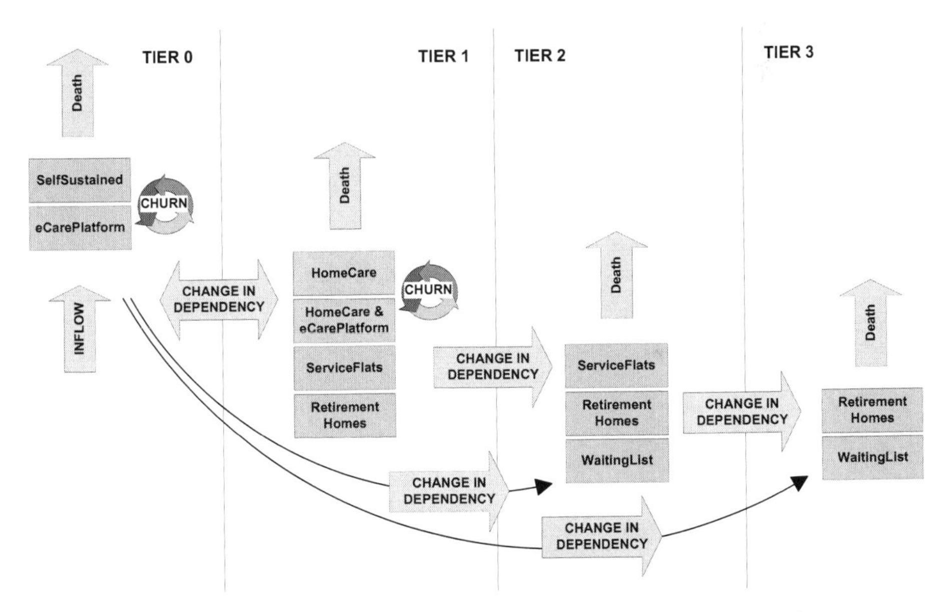


**Fig. 1.** Graphical representation of the developed market share model

### 2.1 Definitions

A number of concepts need to be defined before the model per se can be introduced. First of all, four tiers representing different dependency levels of the elderly are explained. In normal situations, the level of dependency increases on ageing. But at times of life shed events (death of partner, illness, accidents, etc), a deterioration in health can cause suddenly a large increase in dependency.

- **Tier 0** contains those elderly in good condition for whom no extra help is required.
- **Tier 1** elderly require (some) external attention and care, but to such a limited extent that they can stay in their own homes.
- **Tier 2** represents the elderly requiring extended external attention, and for whom staying in their private homes is therefore impossible.
- **Tier 3** elderly require near constant (medical) attention, which only a retirement home can provide.

In these different tiers, competing care providers can be active. The context of these different care providers is defined below. Although not all categories reflect real professional care situations, these definitions allow for uniform modeling of the total envisaged population.

- **SelfSustained (SS)** elderly are those people still living in their private homes which do not need (professional) care.
- **eCarePlatform (eCP)** subscribers do not need (professional) homecare but have chosen to equip themselves with an eCarePlatform to facilitate interaction with informal caregivers (e.g. family) or to subscribe to additional services.
- **HomeCare (HC)** is on-demand supportive care provided in the patient's home by professionals (e.g. health care, cleaning, etc).
- **HomeCare augmented with eCarePlatform (HC+eCP)** is professional homecare facilitated by an eCarePlatform (with more eCare functionalities compared to the platform described above).
- **Service Flats (SF)** or assisted living facilities provide adapted living facilities for elderly who need already a certain level of continued care or assistance.
- **Retirement Homes (RH)** are multi-residence housing facilities intended for highest- level dependency elderly needing around the clock professional care.

Next to the six care providers, we also foresee a **WaitingList (WL)** reflecting the amount of high-dependent elderly who cannot receive adequate care at the time of need, due to capacity limitations in service flats and retirement homes.

## 2.2 General Model Concept

The model is organized around an inflow of customers. A person becomes a part of the scope of the model as soon as he/she turns 60. The model represents 4 tiers of dependency levels numbered 0 to 3. A change in dependency makes the people move to higher tiers. This can happen gradually or a life shed event can make people move from Tier 0 to Tier 2 or even Tier 3. A potential backward move is only possible between Tier 1 and Tier 0 (e.g. after a hospital stay, temporary care can be needed). As customers move into a tier (either from outside the model by turning 60 or from another tier), they are distributed over the available competing providers in this tier. This allocation is based mainly on the current market shares of the providers, corrected by an elasticity (based on offered quality) and limited by capacity. Additionally within Tier 0 and Tier 1, we allow churn (level of customers switching of care provider within the same tier) as people could want to subscribe to the eCarePlatform to bridge certain temporary life situations. Churn is based on a different elasticity and limited by available capacity. People leave a certain tier by moving to a higher level of dependency, or by death. This can be seen in Fig. 1.For a more formal mathematical description of the churn and inflow sub models, please refer to [6].

### 2.3 Inflow and Inter Tier Flow Distribution

The flow distribution system starts from the assumption that customers are allocated according to existing market shares. In other terms if a provider has 30 percent of the market, he should get 30 percent of the inflow of new customers. This would lead to perfectly static market shares. The inflows are therefore modified based on the value offered (perceived quality divided by normalized price) and inflow value elasticity. The inflow value elasticity links the relative change in offered value to the change in percentage of the inflow. An inflow value elasticity of 2 means that if a player offers 10 percent more value than the average of all the providers he should receive 20 percent more customers compared to what he would be entitled to, based purely on his market share. Using this modeling scheme we can benchmark the value offered by a provider by comparing it to the weighted average of the value offered by competition (weighted by market share).

### 2.4 Churn

The churn mechanism is very similar to the flow distribution system. The main difference lies in the definition of the elasticity used. The employed churn value elasticity is defined as the change in market share of a provider relative to the value offered. In other terms a churn value elasticity of 0.5 signifies that if a provider offers a 10 percent better value than the average on the market, his market share should increase with 5 percent. The weighted average value offered has to be used to guarantee that the sum of the churn flow will amount to zero.

### 2.5 Capacity

The capacity model limits the total inflow into a tier to the total available capacity of a tier at that moment in time. This limited flow is then distributed using the inflow and churn model using an iterative process that takes one step less than the number of providers in a tier. With each step (in a capacity limited situation) at least 1 provider is filled up completely by (part of) either his share of the inflow or the churn flow. The remainder is then distributed over the remaining providers of the tier by calculating new flow distributions based on the values offered and the market shares of the remaining competitors. By initially limiting the total inflow to the available capacity, the fact that the customers can always be distributed is assured.

The capacity model calculates capacity for each tier sequentially. It moves from high dependency levels to lower dependency levels (hence from Tier 3 to Tier 0) to give higher priority to high-dependent elderly people. This means that a lack in capacity of a traditional care provider cascades through the model from right to left (see graphical representation in Fig. 1) and can be a contributing factor to an eventual growing in popularity of eHealth solutions for the lower dependency level tiers.

## 3 Simulation Results

We ran simulations based on data for the Flemish area [7][8][9], for a 10 year period (going from 2010 till 2020). The figures illustrate results from our study. However, as the IBBT TranseCare project is still advancing and data is still being collected, refined as well as assessed by professionals from within the Flemish care sector, this might slightly influence the results presented below. Overall it is important to indicate the trend in which competition in the sector will evolve.

### 3.1 Capacity

The total elderly population in Flanders will increase in the upcoming years. This affects the total care sector, which is even today limited by capacity due to a lack of personnel and limitation in subsidized residential elderly care. This problem will furthermore increase in the future. Certainly for high-dependent capacity (service flats and retirement homes), places for new elderly are limited. The waiting lists today are very long [10] and no solution has been proposed, other than to build new service flats and retirement homes.

Fig. 2 shows the outcome from the study indicating the proportion of accepted versus denied capacity (waiting list). As explained in the previous section, people are distributed based on their dependency level in the different tiers (running from high to low dependency). Elderly people desperately in need of help will thus be distributed first. When capacity runs out, people are placed on a waiting list. Note that this is only the case for service flats and retirement homes in Tiers 3 and 2, as we assume that people from Tier 1 have equivalent care providers e.g. homecare. We see that in the beginning, the number of elderly being placed on the waiting list increases due to the current capacity problem in service flats and retirement homes. After year 6, this problem stabilizes at about 7% of people not being able to be helped. This problem is only situated in Tier 2, meaning that people with near constant (medical) attention (Tier 3) will always be helped. The waiting list problem is currently being "solved" by homecare (by professional or informal caregivers). However, homecare augmented with an eCarePlatform could be a (short term) solution for elderly with extended external attention before moving to a service flat or retirement home.

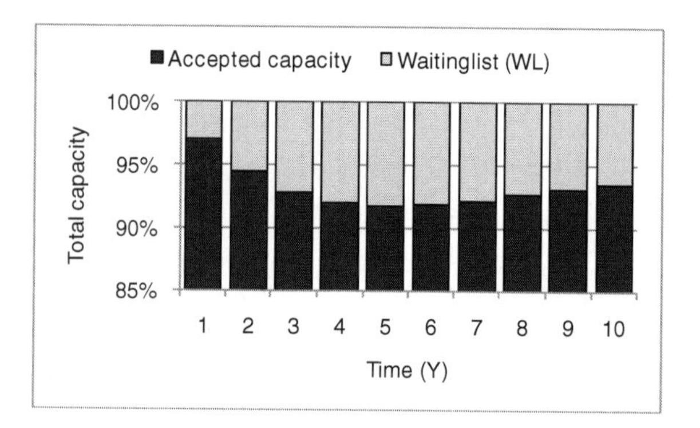


**Fig. 2.** Accepted capacity versus total capacity

## 3.2 Market Shares of Competing Care Providers

When we take a closer look at the market shares of the care providers for the different tiers, we can see the competition between providers offering equal services. The results are presented in Fig. 3 till Fig. 6, showing all care providers per tier (corresponding to Fig. 1), including waiting list (combined for service flats and retirement homes).

Tier 3 results show a full distribution of high-dependency elderly people to retirement homes (RH). The waiting list (WL) remains zero over the time period considered, due to high mortality rate for Tier 3 elderly (24 months passage on average). For Tier 2, all capacity for service flats (SF) is taken up, due to their high value compared to retirement homes (RH). The market share of retirement homes (RH) decreases due to limited capacity available after allocating Tier 3 elderly. Both these issues are reflected in the high waiting list (WL) market share. Solutions could be augmented homecare combined with an eCarePlatform or in worst case hospital admissions when required. In both Tier 1 and Tier 0, two providers compete heavily due to equality in service. In Tier 1 this is homecare with (HC+eCP) and without (HC) eCarePlatform.

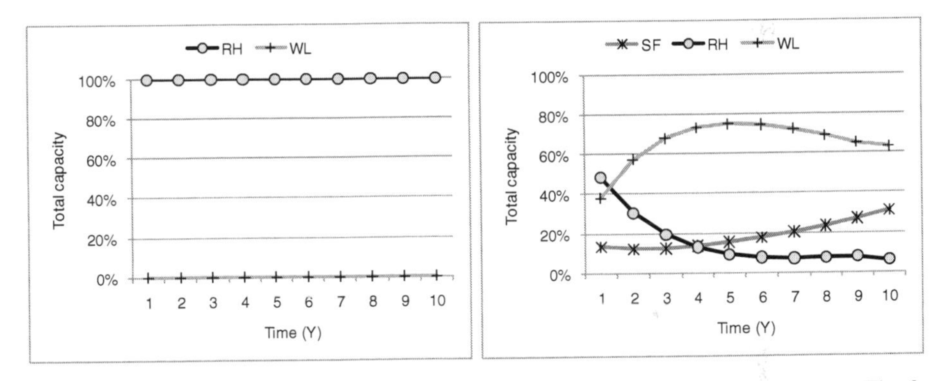


**Fig. 3.** Market shares care providers for Tier 3 **Fig. 4.** Market shares care providers for Tier 2

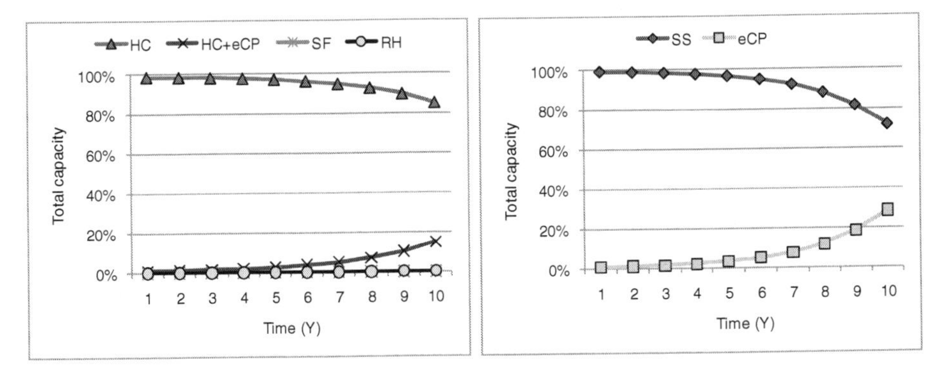


**Fig. 5.** Market shares care providers for Tier 1 **Fig. 6.** Market shares care providers for Tier 0

The latter is considered to have a higher value (due to added value services), which results in the end in a higher market share. The market share of service flats (SF) and retirement homes (RH) is for Tier 1 non-existing due to no available capacity left after allocating Tier 3 and 2 elderly. Home care with or without eCarePlatform is a good alternative. For Tier 0, introducing the eCarePlatform (eCP) could add value to the life of people in good condition (e.g. for solving the problem of social isolation or communicating with family).

The impact of the eCarePlatform could be seen in previous figures. The high value currently used for this provider in our analysis was based on several levels of benefits. First, an active (real time) patient monitoring leads to less medical visits with doctors or professional caretakers, being replaced by online guaranteed assistance. Second on the social level, this medium leads to a better communication with family, care takers, friends, etc. Third on the economic level, this could be a solution to keep control of the overall health care budget and personnel problem in the white sector.

## 4 Conclusion and Future Work

In order to address the increasing costs of providing care to the growing group of elderly, introducing an eCare solution seems an obvious approach. The simulation model presented in this paper allows for forecasting evolution of market shares of competing care providers. Four tiers were defined and presented, based on the dependency level of the elderly, for which the market shares are determined, thereby taking into account available capacity of the different care providers, in- and outflow distribution between tiers and churn between providers within tiers. High-dependency elderly people (Tier 3) could always be allocated to retirement homes. However, there will always be a waiting list for people requiring extended external attention (Tier 2). Augmented home care combined with an eCarePlatform could be a solution to reduce the number of people on the waiting list. For Tier 1 and Tier 0, competition exists between current care providers and providers introducing the eCarePlatform. The latter will gain market share as soon as efficiency and added value has been proven.

This study will be used furthermore as input for the techno-economic analysis of eCare platform business cases, based on eCare value network models. A quantitative study will be performed, making use of multi-actor analysis to calculate and evaluate the business case of each actor involved in the introduction of eHealth services.

## Acknowledgements

This research was carried out as part of the IBBT TranseCare project (https://projects.ibbt.be/transecare/). This project is co-funded by the interdisciplinary research institute IBBT, by the IWT and by Televic, Androme, Custodix, SOL, WGK, In-HAM and UZGent. This work was carried out in the framework of the COST ISO605 Econ@Tel project.

## References

[1] OECD, OECD Regions at a Glance 2009, OECD, Paris (2009)
[2] European Commission. eHealth Benchmarking, Phase II (March 2009), http://ec.europa.eu/information_society/eeurope/i2010/benchmarking/
[3] European Commission. eHealth: a solution for European healthcare systems? The European Files, Nr.17, May-June (2009)
[4] IBBT - Transecare project, https://projects.ibbt.be/transecare/
[5] Van Ooteghem, J., De Maesschalck, S., Bamelis, K., Devos, J., Verhoeve, P., Colle, D., Ackaert, A., Pickavet, M., Demeester, P.: An eHealth business model for independent living systems. In: Proceedings (on CD-ROM) of e-Challenges 2007, The Hague, October 24-26 (2007) ISBN 978-1-58603-801-4
[6] Tesch, T., Descamps, P.T., Van Hoecke, J., Leenknegt, B.: Modeling Market Share Dynamics Using Weighted Averages and Value Elasticities: The M-WAVE Model. In: CTTE 2006 conference proceedings (2006)
[7] FPS Economy - Statistics Division, http://statbel.fgov.be/
[8] Pelfrene, E.: Ontgroening en vergrijzing in Vlaanderen 1990-2050. Ministerie van de Vlaamse Gemeenschap, Administratie Planning en Statistiek (2005)
[9] Eurostat. Demographic Outlook - National reports on the demographic developments in 2007 (January 29, 2009), ISSN 1977-0375
[10] Nieuwsblad. 100.000 wachten op plekje in rusthuis (October 6, 2008), http://www.nieuwsblad.be/Article/Detail.aspx?articleID=qh21cj8m

# Security Protection on Trust Delegated Data in Public Mobile Networks

Dasun Weerasinghe, Muttukrishnan Rajarajan, and Veselin Rakocevic

School of Engineering and Mathematical Sciences
City University London
dasun.weerasinghe@city.ac.uk

**Abstract.** This paper provides detailed solutions for trust delegation and security protection for medical records in public mobile communication networks. The solutions presented in this paper enable the development of software for mobile devices that can be used by emergency medical units in urgent need of sensitive personal information about unconscious patients. In today's world, technical improvements in mobile communication systems mean that users can expect to have access to data at any time regardless of their location. This paper presents a token-based procedure for the data security at a mobile device and delegation of trust between a requesting mobile unit and secure medical data storage. The data security at the mobile device is enabled using identity based key generation methodology.

## 1 Introduction

In the modern world people are getting used to have access to a wide range of data and applications wherever they are and whenever they want, using public mobile communication networks. In such a ubiquitous communication environment, it is not a surprise that there is a growing need to enable the emergency medical teams to have a continuous and secure access to patient medical records. The added benefit of having person's medical record while providing emergency care is obvious, and has been highlighted in a number of publications [1][7][6].

This paper provides a detailed analysis of the above issues and provides solutions for secure authentication of data download and data protection following the download. In this respect, the first issue that requires attention is the secure authentication. This is achieved by careful distribution of trust between the key players in the process: the medical provider storing the medical records, the mobile network, and the mobile device requesting the data. Trust had to be negotiated and delegated between these players to enable them to feel confident to exchange data. The authors have presented the trust negotiation methodology in one of their previous publications[10].

With the emergence of electronic health solutions, the delegation and negotiation of trust from one healthcare service provider (HSP) to another is one of the main requirements for the secure provision of data and services [9]. The healthcare service providers can "in the extreme case" be mutually unknown and therefore not trusting each other. In our paper, the HSPs are classified in two categories; the 'relying healthcare service provider' and the 'requesting healthcare service provider'. The relying HSP is a medical

P. Kostkova (Ed.): eHealth 2009, LNICST 27, pp. 200–207, 2010.

center or a hospital which stores sensitive patient medical records - including patient's medical history, current diagnosis and medical treatments, known allergies, social history of the patient and patient personal information. The patients have the ownership of the patient medical records but they have granted the trust delegation on accessing these records to their HSP [8]. The requesting HSP is another medical center, hospital or mobile healthcare service unit with doctors and/or paramedics. This HSP requests access to patient medical records from the relying HSP in order to perform special or urgent diagnosis and medical treatment to patients. The access to the patient medical record is vital for a doctor at the requesting HSP to perform a correct diagnosis and/or treatment. This paper provides a detailed solution for securing this scenario.

## 2 Trust Negotiation in Mobile Services

During the recent past, initiatives have been taken both by the academia and by the industries towards improving the use of mobile communication for healthcare and safety of the public [4]. The m-health is an existing term representing an emerging set of healthcare applications and services that people can access from their web-enabled mobile devices [2]. Medical personnel having access to clinical data irrespective of the geographic location is an advantage of m-health. There are numerous examples of interesting applications. For example, real-time mobile telemedicine system is introduced to transmit video and patient bio-signals from a moving ambulance to a doctor in the hospital using wireless cellular phones [12]. Mobile device in the ambulance is connected to a Web service in the hospital to retrieve advices about transferring the patient there [5]. These approaches allow medical personnel to access patient medical records from a remote location but only if the patient medical records are at a centralized or distributed location for public access. Generally patient medical records are stored at patient's medical center and access to those records are restricted to protect the data confidentiality and patient privacy. Therefore mobile medical personnel at the disaster scene has to prove the legitimacy to access patient medical records from the patient's medical center [8].

A trust negotiation process should incorporate a trust negation algorithm to identify, verify and validate the trust level of the requestor party with respect to the requesting information. There are number of trust negotiation algorithms available and our framework will be able to use any of those to generate the trust level between the requestor party and relying party. Wu Z. et al describes an indirect trust establishment mechanism to bridge and build new trust relationships from extant trust relationships [11]. The trust evaluation algorithms output a trust level defined in the rage of Full to Minimal such as Full, High, Medium, Low, Minimal or the rage is in numerical numbers such as 1 to 10 [3].

Transferring trust delegation for accessing patient medical records between healthcare service providers is one of the vital requirements in healthcare industry and specially accessing patient medical records over a mobile device in emergency situations. According to the knowledge of authors most publications haven't considered the security and privacy aspects in trust negotiation techniques for mobile healthcare environment. Therefore the novelty and the research contribution of this paper compared to the other publications is; 'Token based trust negotiation and delegation framework for healthcare service providers with a security and privacy protection on trust delegated data'.

## 3 Proposed Schema

The solution for trust negotiation in mobile web services is designed using the token based trust negotiation framework. The TGS is the facilitator for the trust negotiation between healthcare service providers. It generates and issues tokens for authentication and trust negotiation process. These tokens are designed in XML format and those are categorized into security tokens and trust tokens. The mobile device is unlikely to be trusted by the schema but the security capsule is a trusted entity. So patient medical records and obtained tokens are stored in the security capsule. The patient medical records are stored in the encrypted format and security capsule can decrypt those only if valid security and trust tokens are present.

The use case for trust delegation on patient medical records begins when MHP attempts to access patient medical record from a healthcare service provider. The patient medical records are saved at the relying healthcare service provider and TGS bridges the trust negotiation between two parties. The scheme for transferring trust delegation to access patient medical records is summarized with the reference to Figure 1.

1. The MHP authenticates with TGS to access patient medical records. The TGS issues the security token to the MHP's mobile device
2. The MHP requests the access to patient medical record from TGS by specifying the patient identity
3. The TGS locates the relying healthcare service provider for the patient medical records and sends the trust negotiation request
4. The patient medical record and trust tokens are sent to the mobile device of MHP from the patient's healthcare service provider
5. The mobile device decrypts the patient medical records utilizing the tokens records.

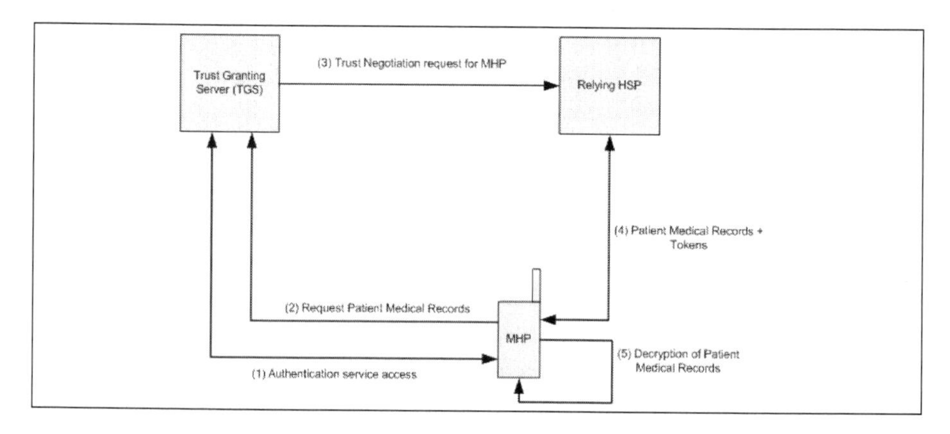


**Fig. 1.** Scheme Description

# 4 Implementation

The paper presents both specific protocol exchanges and the structure and syntax of security and trust tokens.

## 4.1 Protocol

This section describes the critical protocol exchanges to address the threat model with the consideration of authentication, confidentiality and integrity. The protocol consists of 2 phases:

1. MHP authenticates with TGS
2. Trust negotiation between relying healthcare service provider and MHP

The following additional notations are adapted for the protocol explanation:

- RelHSP= Relying healthcare service provider
- ReqHSP= Requesting healthcare service provider

**Phase 1: MHP authenticates with the TGS**
Phase 1 initiates with MHP going to a disaster scene. The mobile device of MHP had the Login Token that was generated by the healthcare service provider. Following are the steps to get the MHP authenticated with the TGS:

1. MHP to TGS [Login Token]; The login token is the authentication request to the TGS from MHP. The token consists of information about the healthcare service provider and mobile healthcare personal.
2. The TGS decrypts the message using its private key and verifies the signature of the token against the public key certificate of ReqHSP. If the verification is successful then the ReqHSP and MHP identification are checked in the Trust Mapping Database.
3. TGS to MHP [Authentication Token]; Once the trust level is obtained, the TGS generates the Authentication Token for MHP.

**Phase 2: Trust Negotiation between MHP and Relying Healthcare Service provider**
Phase 2 startes with MHP approaching a patient at a disaster situation. The patient needs urgent medical attention and MHP has to view the patient medical records for effective treatments. It is assumed that MHP has found an identification of the patient.

Following are the steps on trust negotiation between the healthcare service provider and the MHP:

1. MHP to TGS: [RecordAccess(PatientID, Authentication Token)]; The MHP identifies the patient and makes the request to access patient medical records.
2. The TGS verifies the authentication token and then identifies the relying healthcare service provider (RelHSP) of the patient. The RelHSP holds the patient medical records. Then TGS locates the previous trust negotiation and trust decline records between the requesting parties (MHP and ReqHSP) and the relying party (RelHSP) in the trust mapping database. The Trust Evaluation Engine generates the recommended trust level for MHP to access patient medical records from RelHSP.

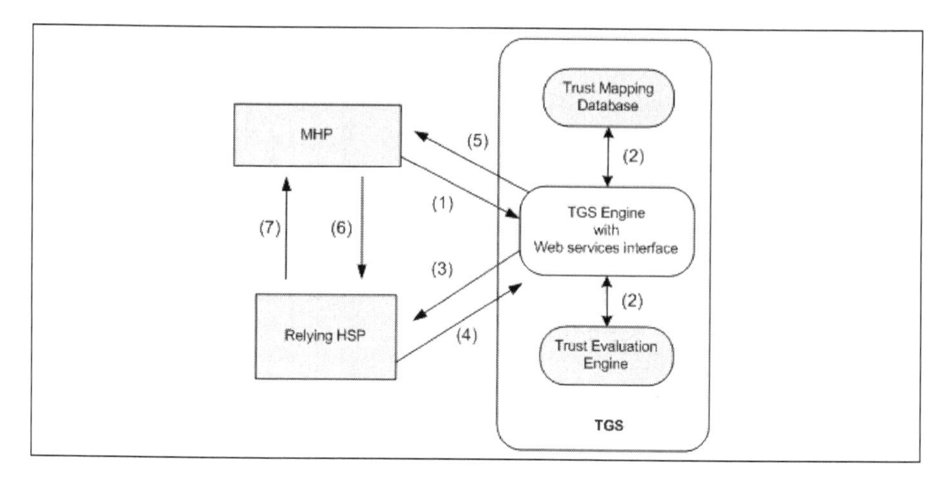


**Fig. 2.** Phase 2 message flow

3. TGS to RelHSP [TrustRecommendation Token]; The Trust Recommendation Token is generated by the TGS.
4. RelHSP to TGS [TrustChallenge, Confirmation, Assigned Trust Level]; The RelHSP verifies the Trust Recommendation Token against the TGS signature and obtains the information about the requesting party and patient identification. The finalized trust level for MHP is named as 'Assigned Trust Level' and then the trust challenge token is generated by RelHSP.
5. TGS to MHP [TrustChallenge Token]. The TGS sends the Trust Challenge Token to the MHP.
6. MHP to RelHSP [TrustChallengeResponse]. The MHP retrieves the RAND from the trust challenge token and generates the trust challenge response using its private key for integrity.
7. RelHSP to MHP [Trust Token]. The RelHSP validates the trust challenge response against the RAND and public key certificate of MHP. If the validation is successful then the Trust Token is generated as trust delegation object.

Finally the RelHSP encrypts the patient medical records using the session key of the trust token and transmits it to the MHP. The data is signed by the RelHSP private key for integrity and encrypted by the MHP public key for confidentiality.

### 4.2 Token Generation and Management

This section describes the token structures for the proposed schemas and the below abbreviations are used for token representation.

$$TS = \text{Time stamp}$$
$$s_{N_K}(\text{X}) = \text{The signature of data X using secret key K of N}$$
$$e_{N_K}(\text{X}) = \text{The encryption of data X using secret key K of N}$$

- Authentication Token (AT)
  ( AT = $e_{TGS_{S1}}(s_{TGS_{S2}}$[ ReqHSPID | MHPID | GTL | Token Life Time | TS ]));
  The Authentication Token is issues by the TGS for authenticated mobile healthcare personals. This token is belonged to the TGS and this can only be viewed and verified by the TGS. Therefore the token is signed by TGS integrity key (TGS2) and then encrypted by the confidentiality key of TGS (TGS1). This token specifies the General Trust Level (GTL) of the MHP.

- Trust Token(TT)
  ( TT = $e_{MHP_{public}}(s_{RelHSP_{private}}$[ TTID | MHPID | ATL | PatientID | Token Life Time | TS | tsK ]));
  The trust token is the trust granted object for MHP to access the requested patient medical record. The trust token identification (TTID) is assigned to each trust token for unique identification. The tsK is the session key is used to decrypt the encrypted patient medical records.

### 4.3 Security Capsule Implementation

The security capsule has been developed to enable security of the healthcare data in a mobile device and it is a application for a mobile device that can be installed using OTA technique. The patient medical records are sent to the mobile device in encrypted format and the encrypted data is saved in the security capsule. The data decryption and protect the security and privacy on patient medical records are the main functionalities of it.

The logical architecture of the security capsule consists of six functional and storage units as shown in Figure 3. The Service Manager establishes Web services communication with the Trust Granting Server and Healthcare Service providers. It establishes the communication with the external parties. Meanwhile it communicates with the mobile device display API and data storage units. The Data Manager, Token Manager and Key Manager maintain storage spaces respectively for encrypted data, tokens and cryptographic keys. The Service Manager filters the incoming data stream and then dispatch it to the correct storage area. The incoming encrypted medical records from the service providers are decrypted by the process manager.

The data decryption process executes when the mobile personal requests to view the data from the mobile device. Then the Process Manager retrieves the encrypted data from the Data Manager and the relevant tokens from the Token Manager storage. The token validation and public key certificate validation functions are performed to verify the legitimacy of the tokens and cryptographic keys. Finally the Process Manager generates the decryption key and then decrypts the encrypted data. The decrypted data is displayed in the mobile device through the Service Manager and after the session the decrypted data and cryptographic keys are sent to the Data Dump Manager. The Data Dump Manger discards used data, tokens and keys from the device and USIM memory. The decryption key generation process and the permanent data deletion process in the security capsule protect the patient medical records from privacy and security vulnerabilities.

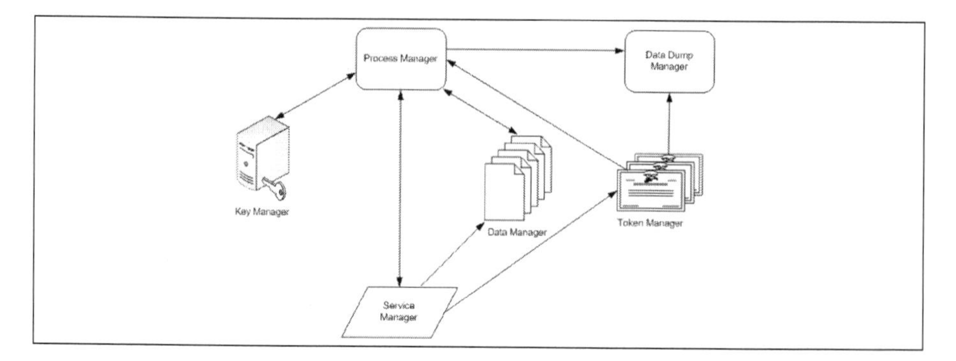


**Fig. 3.** Security Capsule Architecture

The novel key generation algorithm in the Process Manger generates the data key to decrypt encrypted patient medical records from the service providers. Following are the input types for the key generation algorithm.

- User PIN: 4 digit PIN agreed between mobile healthcare personal and the healthcare service provider
- IMPI (IP Multimedia Private Identity): The mobile operator assigned identity for the mobile healthcare personal with the USIM
- IMEI (International Mobile Equipment Identity): The unique identity for the mobile device and this is issued by the mobile device manufacturer.
- UID: The Identity provider issued unique identity for the security capsule
- Token Key: The cryptographic key in the token that provides mobile device authentication to the decrypt data
- Vendor Key: the cryptographic key is sent by the healthcare service provider during the decryption process. This key provides a real time authentication of the mobile healthcare personal with the User PIN.

Therefore the mobile device dependent, mobile SIM dependent, Mobile personal dependent, TGS dependent and the Healthcare service provider dependent identification parameters are required in the key generation process. This will protect transmitting patient medical records to other mobile devices or stealing mobile devices to access sensitive data inside.

## 5 Conclusion

The paper has introduced a scheme for the trust negotiation between healthcare service providers to retrieve and access patient medical records using mobile devices during an emergency scene. The main contribution of the paper can be summarized as; 'Trust negotiation and data protection at a mobile device'. The contribution of system architecture, scheme and protocol will form a new business model for healthcare industry to efficiently and securely share data and services between unknown healthcare service providers.

## References

1. Belsis, M.A.: Dwivedi A.N. Providing secure maccess to medical information. Int. J. Electronic Healthcare 3(1), 51–57 (2007)
2. Istepanian, R.S.H., Jovanov, E., Zhang, Y.T.: Guest editorial introduction to the special section on m-health: Beyond seamless mobility and global wireless health-care connectivity. IEEE Transactions on Information Technology in Biomedicine 8(4), 405–414 (2004)
3. Jung, D.I., Avolio, B.J.: Opening the black box: an experimental investigation of the mediating effects of trust and value congruence on transformational and transactional leadership. Journal of Organizational Behavior 21(8), 949–964 (2000)
4. Moran, E.B., Tentori, M., Gonzalez, V.M., Favela, J., Martinez-Garcia, A.I.: Mobility in hospital work: towards a pervasive computing hospital environment. International Journal of Electronic Healthcare 3(1), 72–89 (2007)
5. Motta, E., Domingue, J., Cabral, L., Gaspari, M.: Irs-ii: A framework and infrastructure for semantic web services. In: Fensel, D., Sycara, K., Mylopoulos, J. (eds.) ISWC 2003. LNCS, vol. 2870, pp. 306–318. Springer, Heidelberg (2003)
6. Mu, M.A., Rodrguez, M., Favela, J., Martinez-Garcia, A.I., Gonzlez, V.M.: Context-aware mobile communication in hospitals. Computer 36(9), 38–46 (2003)
7. Rodriguez, M.D., Favela, J., Martinez, E.A., Munoz, M.A.: Location-aware access to hospital information and services. IEEE Transactions on Information Technology in Biomedicine 8(4), 448–455 (2004)
8. Schoenberg, R., Safran, C.: Internet based repository of medical records that retains patient confidentiality. British Med. J. 321, 1199–1203 (2000)
9. Vawdrey, D.K., Sundelin, T.L., Seamons, K.E., Knutson, C.D.: Trust negotiation for authentication and authorization in healthcare information systems. In: Proceedings of the 25th Annual International Conference of the IEEE (September 2003)
10. Weerasinghe, D., Rajarajan, M., Rakocevic, V.: Trust delegation for medical records access using public mobile networks. In: Proceedings of the 3rd International Conference on Pervasive Computing Technologies for Healthcare (April 2009)
11. Wu, Z., Weaver, A.C.: Bridging trust relationships with web service enhancements. In: ICWS 2006: Proceedings of the IEEE International Conference on Web Services, pp. 163–169 (2006)
12. Xiao, Y., Gagliano, D., LaMonte, M., Hu, P., Gaasch, W., Gunawadane, R., Mackenzie, C.: Design and evaluation of a real-time mobile telemedicine system for ambulance transport. J. High Speed Netw. 9(1), 47–56 (2000)

# Communicating with Public Health Organizations: Technical Solution

Alexandru Mihai*, Daniel Catalan, Skaidra Kurapkiene, and Wadih Felfly

European Centre for Disease Prevention and Control, Director's Cabinet,
Tomtebodavagen 11A, SE-171 83, Stockholm, Sweden
{alexandru.mihai,daniel.catalan,skaidra.kurapkiene,
wadih.felfly}@ecdc.europa.eu

**Abstract.** By working with experts throughout Europe, ECDC pools Europe's health knowledge, so as to develop authoritative scientific opinions about the risks posed by current and emerging infectious diseases. Difficulties rose in the management of competent bodies' lists and the information was duplicated several times across the organization. ECDC started implementing a CRM system to organize the information in a structured data model, track the history of communication, provide contact information to application in house and support the nomination process and the user identity management for these applications.

**Keywords:** CRM, information management, public health.

## 1 Introduction and Requirements

The European Centre of Disease Prevention and Control (ECDC) is an EU agency established in 2005 with the mission to identify, assess and communicate current and emerging threats to human health posed by infection diseases. In order to achieve this mission, ECDC works in partnership with national protection bodies across Europe to strengthen and develop continent-wide disease surveillance and early warning systems. By working with experts throughout Europe, ECDC pools Europe's health knowledge, so as to develop authoritative scientific opinions about the risks posed by current and emerging infectious diseases.

ECDC activities rely on coordination and constantly involve communication and exchange of information with many external organizations and people. ECDC is a fast growing organization, both in terms of people and functional coverage and activities.

Difficulties are arising in the management of contact lists and the information is duplicated several times across the organization. Many files are outdated while contacting the right person having the right role is critical in the daily activity of ECDC.

---

* Please note that the LNICST Editorial assumes that all authors have used the western naming convention, with given names preceding surnames. This determines the structure of the names in the running heads and the author index.

P. Kostkova (Ed.): eHealth 2009, LNICST 27, pp. 208–209, 2010.

To support its growth, ECDC implemented a Microsoft Dynamics CRM 4.0 system to organize the information about external entities in a structured data model.

In many cases, ECDC relies on past events and the corresponding actions that were taken to build up its knowledge base and experience in order to speed up response to present and future threats. Keeping the history of communications and people contacted regarding a subject becomes an important part of the global knowledge base strategy.

The aim of the CRM system is to provide ECDC with a tool to help structure information about countries, organizations, people, documents and activities. Many projects were identified across ECDC organization, some of which can be integrated and interfaced to work seamlessly with the CRM.

The target is to put the CRM as the main system for contacts management. CRM will interface with other systems and provide them with contacts management features and services.

## 2 Implementation Phases

To accelerate the delivery of the CRM, the solution will be implemented in several phases. High priority functionalities are implemented in early phases.

**Phase 1** was composed of the core CRM functions requested by all units: structure information about countries, organizations and people; organize and manage contact lists in a standard format across ECDC; easily find competent bodies and the designated contacts; categorize contacts and create groups of contacts (related to an organization, network, project, topic…); display organizations by country, and contacts by organization or by group; search and find entities using multiple criteria; manage meetings and visits; save the history of the exchanges with contacts; print, sort and export information; mail merge with Microsoft office documents; integration in Outlook; possibility of attaching files to entities; access the information remotely or while not connected to ECDC network.

**Phase 2** is composed of the unit specific CRM functions: customized templates for all units; automates creating badges and documents for meetings, personalized letters and emails; customized views for all units; implementation of hidden fields by unit for CAB and HCU; marketing module for HCU: mailing lists and communication campaigns; interface with EU publication office for HCU; service module for PRU: cases to group outbreak related communication; interface with document management server for SAU.

**Phase 3** is composed mainly of interfaces with other systems and ongoing projects: interface with Portal; interface with Active directory; interface with EWRS; interface with Experts database; interface with terminology server; interface with SAP.

**Phase 4** is composed of optional projects and requests: articles management and review for HCU; customization of service module for crisis management for PRU; contract management for ADMIN; interface with registry system for CAB/ADMIN; interface with EPIS; integration of EPIET.

# Author Index